Peer Review in Environmental Technology Development Programs

THE DEPARTMENT OF ENERGY'S OFFICE OF SCIENCE AND TECHNOLOGY

Committee on the Department of Energy–Office of Science and
Technology's Peer Review Program
Board on Radioactive Waste Management
Commission on Geosciences, Environment, and Resources
National Research Council

NATIONAL ACADEMY PRESS
Washington, D.C. 1998

NOTICE: The project that is the subject of this report was approved by the Governing Board of the National Research Council, whose members are drawn from the councils of the National Academy of Sciences, the National Academy of Engineering, and the Institute of Medicine. The members of the committee responsible for the report were chosen for their special competences and with regard for appropriate balance.

This work was sponsored by the U.S. Department of Energy, Contract No. DE-FC01-94EW54069. All opinions, findings, conclusions, and recommendations expressed herein are those of the authors and do not necessarily reflect the views of the Department of Energy.

International Standard Book Number 0-309-06338-8

Additional copies of this report are available from:

National Academy Press
2101 Constitution Ave., NW
Box 285
Washington, DC 20055
800-624-6242
202-334-3313 (in the Washington Metropolitan Area)
http://www.nap.edu

Copyright 1998 by the National Academy of Sciences. All rights reserved.

Printed in the United States of America

COMMITTEE ON THE DEPARTMENT OF ENERGY–OFFICE OF SCIENCE AND TECHNOLOGY'S PEER REVIEW PROGRAM

C. HERB WARD, *Chair,* Rice University, Houston, Texas
BARRY BOZEMAN, Georgia Institute of Technology, Atlanta
RADFORD BYERLY, JR., University Corporation for Atmospheric Research (retired), Boulder, Colorado
LINDA A. CAPUANO, AlliedSignal Aerospace, San Jose, California
RICHARD A. CONWAY, Union Carbide Corporation (retired), South Charleston, West Virginia
THOMAS A. COTTON, JK Research Associates, Vienna, Virginia
FRANK P. CRIMI, Lockheed Martin Advanced Environmental Systems Company (retired), Saratoga, California
JOHN C. FOUNTAIN, State University of New York, Buffalo
DAVID T. KINGSBURY, Chiron Corporation, Emeryville, California
GARETH THOMAS, University of California, Berkeley

Staff

GREGORY H. SYMMES, Study Director[*]
SUSAN B. MOCKLER, Research Associate
ERIKA L. WILLIAMS, Research Assistant
ROBIN L. ALLEN, Senior Project Assistant

---

[*]Commission on Geosciences, Environment, and Resources.

BOARD ON RADIOACTIVE WASTE MANAGEMENT

MICHAEL C. KAVANAUGH, *Chair,* Malcolm Pirnie, Inc., Oakland, California
JOHN F. AHEARNE, *Vice-Chair*, Sigma Xi, The Scientific Research Society, and Duke University, Research Triangle Park and Durham, North Carolina
ROBERT J. BUDNITZ, Future Resources Associates, Inc., Berkeley, California
ANDREW P. CAPUTO, Natural Resources Defense Council, Washington, D.C.
MARY R. ENGLISH, The University of Tennessee, Knoxville
DARLEANE C. HOFFMAN, Lawrence Berkeley Laboratory, Berkeley, California
JAMES H. JOHNSON, JR., Howard University, Washington, D.C.
ROGER E. KASPERSON, Clark University, Worcester, Massachusetts
JAMES O. LECKIE, Stanford University, Stanford, California
JANE C.S. LONG, University of Nevada, Reno
CHARLES McCOMBIE, NAGRA, Wettingen, Switzerland
ROBERT MEYER, Keystone Scientific, Inc., Fort Collins, Colorado
WILLIAM A. MILLS, Oak Ridge Associated Universities, (retired), Olney, Maryland
D. WARNER NORTH, NorthWorks, Inc., Mountain View, California
MARTIN J. STEINDLER, Argonne National Laboratory, Argonne, Illinois
JOHN J. TAYLOR, Electric Power Research Institute, Palo Alto, California
MARY LOU ZOBACK, U.S. Geological Survey, Menlo Park, California

NRC Staff

KEVIN D. CROWLEY, Director
ROBERT S. ANDREWS, Senior Staff Officer
THOMAS E. KIESS, Senior Staff Officer
JOHN R. WILEY, Senior Staff Officer
SUSAN B. MOCKLER, Research Associate
ERIKA L. WILLIAMS, Research Assistant
TONI GREENLEAF, Administrative Associate
ROBIN L. ALLEN, Senior Project Assistant
PATRICIA A. JONES, Senior Project Assistant
ANGELA R. TAYLOR, Senior Project Assistant
LATRICIA C. BAILEY, Project Assistant
LAURA LLANOS, Project Assistant

COMMISSION ON GEOSCIENCES, ENVIRONMENT, AND RESOURCES

GEORGE M. HORNBERGER, *Chair*, University of Virginia, Charlottesville
PATRICK R. ATKINS, Aluminum Company of America, Pittsburgh, Pennsylvania
JERRY F. FRANKLIN, University of Washington, Seattle
B. JOHN GARRICK, PLG, Inc., Newport Beach, California
THOMAS E. GRAEDEL, Yale University, New Haven, Connecticut
DEBRA KNOPMAN, Progressive Policy Institute, Washington, D.C.
KAI N. LEE, Williams College, Williamstown, Massachusetts
JUDITH E. McDOWELL, Woods Hole Oceanographic Institution, Woods Hole, Massachusetts
RICHARD A. MESERVE, Covington & Burling, Washington, D.C.
HUGH C. MORRIS, Canadian Global Change Program, Delta, British Columbia
RAYMOND A. PRICE, Queen's University at Kingston, Ontario
THOMAS C. SCHELLING, University of Maryland, College Park
VICTORIA J. TSCHINKEL, Landers and Parsons, Tallahassee, Florida
E-AN ZEN, University of Maryland, College Park
MARY LOU ZOBACK, United States Geological Survey, Menlo Park, California

NRC Staff

ROBERT M. HAMILTON, Executive Director
GREGORY H. SYMMES, Assistant Executive Director
JEANETTE SPOON, Administrative Officer & Financial Officer
SANDI FITZPATRICK, Administrative Associate
MARQUITA SMITH, Administrative Assistant/Technology Analyst

The National Academy of Sciences is a private, nonprofit, self-perpetuating society of distinguished scholars engaged in scientific and engineering research, dedicated to the furtherance of science and technology and to their use for the general welfare. Upon the authority of the charter granted to it by the Congress in 1863, the Academy has a mandate that requires it to advise the federal government on scientific and technical matters. Dr. Bruce M. Alberts is president of the National Academy of Sciences.

The National Academy of Engineering was established in 1964, under the charter of the National Academy of Sciences, as a parallel organization of outstanding engineers. It is autonomous in its administration and in the selection of its members, sharing with the National Academy of Sciences the responsibility for advising the federal government. The National Academy of Engineering also sponsors engineering programs aimed at meeting national needs, encourages education and research, and recognizes the superior achievements of engineers. Dr. William A. Wulf is president of the National Academy of Engineering.

The Institute of Medicine was established in 1970 by the National Academy of Sciences to secure the services of eminent members of appropriate professions in the examination of policy matters pertaining to the health of the public. The Institute acts under the responsibility given to the National Academy of Sciences by its congressional charter to be an adviser to the federal government, and, upon its own initiative, to identify issues of medical care, research, and education. Dr. Kenneth Shine is president of the Institute of Medicine.

The National Research Council was organized by the National Academy of Sciences in 1916 to associate the broad community of science and technology with the Academy's purposes of furthering knowledge and advising the federal government. Functioning in accordance with general policies determined by the Academy, the Council has become the principal operating agency of both the National Academy of Sciences and the National Academy of Engineering in providing services to the government, the public, and the scientific and engineering communities. The Council is administered jointly by both Academies and the Institute of Medicine. Dr. Bruce M. Alberts and Dr. William A. Wulf are chairman and vice-chairman, respectively, of the National Research Council.

Preface

In March 1996, the National Research Council's Committee on Environmental Management Technologies (CEMT) recommended that the Department of Energy (DOE)–Office of Science and Technology (OST) develop and apply a standardized, rigorous, and independent external peer review process to all of its technology development programs (NRC, 1996). A subsequent report by the General Accounting Office (GAO, 1996) echoed these findings. As a result, in September 1996, OST instituted a new program to perform peer reviews of technologies (or groups of technologies) at various stages of development. Shortly thereafter, OST requested that the National Research Council (NRC) form a committee to evaluate the effectiveness of its new program and make specific recommendations to improve it, if appropriate. In particular, the committee was asked to review the following:

- internal procedures used by OST to identify the need for timely peer review of projects and programs;
- structures, protocols, and procedures for obtaining peer reviews of OST projects and programs, including who decides what will be peer reviewed, what criteria for peer review are used, and when in the R&D process peer review is requested; and
- feedback of peer review results into program management and development decisions.

The committee was directed to compare OST's practices to generally accepted norms for scientific and technical peer review, including practices for selection of peer reviewers and screening for bias and conflict of interest.

This is the second of two reports prepared by the committee. OST requested an interim report (NRC, 1997b) to provide a preliminary assessment of the program. In particular, OST asked the committee to consider whether it was moving in the right direction toward the implementation of a credible, effective, and defensible peer review program. In this final report, the committee provides a more detailed assessment of OST's peer review program after its first complete annual cycle.

Another NRC committee, the Committee on Prioritization and Decision Making in the Department of Energy–Office of Science and Technology, is currently conducting a parallel evaluation of the decision-making processes throughout OST's technology development program. One aspect of its work will be to examine the role and importance of peer reviews in OST's decision-making processes. Our committee therefore has focused its work on OST's peer review program itself (including an evaluation of how peer reviews, if conducted, could be made more useful as an input to OST's decision-making processes), but has not evaluated these decision-making processes explicitly.

In conducting this study, the committee was briefed on the newly instituted peer review program by OST staff at six committee meetings. The committee wishes to thank Gerald Boyd, Acting Deputy Assistant Secretary for Science and Technology, in particular, for meeting with the committee at three of its meetings. In addition, Jef Walker, Anibal Taboas, Texas Chee, Miles Dionisio, and other OST staff were very helpful in providing requested materials; briefing the committee on various aspects of the peer review program; sharing various "lessons learned" about the peer review program; and coordinating briefings by other relevant DOE staff. They also helped facilitate committee members' observations of a number of the peer reviews. The committee also would like to thank the staff of the Institute for Regulatory Science (RSI) and representatives from the American Society of Mechanical Engineers (ASME)—most notably, Alan Moghissi, Ernest Daman, and Howard Clark—for arranging for the committee to observe selected peer reviews, providing the committee with draft peer review reports, and sharing their thoughts on OST's peer review program.

In addition to briefings by DOE, RSI, and ASME, the committee also heard from representatives from other organizations that utilize and/or conduct peer reviews. The committee wishes to thank all of the invited speakers who made presentations to the committee on peer review practices in their organizations, specifically, Brad Smith from the Department of Defense's Strategic Environmental Research and Development Program, Donna Dean from the National Institutes of Health, Hratch Semerjian from the National Institute of Standards and Technology, Donald Senich from the National Science Foundations' Small Business Innovation Research Program, Dorothy Patton and Jack Puzak from the Environmental Protection Agency (EPA), David Morrison from the U.S. Nuclear Regulatory Commission, Don Barnes from EPA's Science

Advisory Board, Robert Marianelli from the Department of Energy–Office of Basic Energy Sciences, Donald R. Beem from the American Institute of Biological Sciences, and Carl Guastaferro from Information Dynamics, Inc. The latter two organizations have conducted reviews for federal agencies for many years. Although the mission of each of these organizations is distinct from that of OST, these presentations illustrated that many of the fundamental characteristics of successful peer review programs are applicable to a diverse range of objectives, from reviewing proposals to reviewing ongoing technology projects.

This report has been reviewed in draft form by individuals chosen for their diverse perspectives and technical expertise, in accordance with procedures approved by the NRC's Report Review Committee. The purpose of this independent review is to provide candid and critical comments that will assist the institution in making the published report as sound as possible and to ensure that the report meets institutional standards for objectivity, evidence, and responsiveness to the study charge. The review comments and draft manuscript remain confidential to protect the integrity of the deliberative process. We wish to thank the following individuals for their participation in the review of this report:

> Paul Barton, U.S. Geological Survey (emeritus)
> Daryl Chubin, National Science Foundation
> Mary English, University of Tennessee
> George Hornberger, University of Virginia
> James Johnson, Howard University
> Jeff Marqusee, U.S. Department of Defense
> John Taylor, Electric Power Research Institute (retired)

While the individuals listed above have provided constructive comments and suggestions, it must be emphasized that responsibility for the final content of this report rests entirely with the authoring committee and the institution.

This report could not have been completed without the able assistance of National Research Council staff. Robin Allen provided meeting and committee support during the early stages of this study, as well as assisting with the final review and publication of this report. Rob Greenway prepared the camera-ready copy of this report for publication. Susan Mockler prepared meeting minutes, conducted research, and edited several drafts of the report. Erika Williams provided exceptional meeting and committee support for most of the study, and made significant contributions to the report by compiling committee members' written contributions into coherent drafts and by conducting research. The committee is especially grateful to study director

Gregory Symmes, whose writing skills and insights contributed significantly to this report.

One of the challenges of this study has been that OST's peer review program has been somewhat of a "moving target" during the study period, because OST has made a number of changes in the program throughout the study, especially in response to the committee's interim report. Although this has made the committee's work more difficult at times, it reflects a positive commitment within OST to improve the peer review program as potential improvements are recognized. The committee hopes that this report will be useful to OST as it strives to implement peer review as an important tool in its decision-making process.

C. Herb Ward, Chair

Contents

EXECUTIVE SUMMARY   1

1   INTRODUCTION   21

   OST's Peer Review Program, 22
   Charge to the Committee, 23
   Parallel Efforts, 24
   Study Process, 24
   Reports, 25

2   ELEMENTS OF A CREDIBLE PEER REVIEW PROGRAM   27

   Definition of Peer Review, 27
   Benefits of Peer Review, 29
   Peer Review Process, 31

3   USES OF PEER REVIEW   38

   Proposal Evaluations, 39
   Project Maturity Evaluations, 41
   Program Balance Evaluations, 46
   Needs Determinations, 47

4   OST'S PEER REVIEW PROGRAM   53

   Technology Investment Decision Model, 53

Roles and Responsibilities, 54
Selection of Technologies to Be Reviewed, 60
Documentation Required for Review, 60
Types of Reviews, 61
Selection of Reviewers, 62
Review Criteria, 63
Review Session, 64
Review Reports, 65
Feedback of Peer Review Results into Program Management and Development Decisions, 66

5   ANALYSIS OF OST'S PEER REVIEW PROGRAM   67

Definition of Peer Review, 68
Benefits of Peer Review, 68
Peer Review Process, 69

6   "TRIAGE" APPROACH FOR REDUCING PROJECT BACKLOGS   82

7   IMPROVING THE EFFECTIVENESS OF OST'S PEER REVIEW PROGRAM   85

Linkage of Peer Reviews to Management Decisions, 86
Evaluation and Improvement Mechanisms, 87
Potential Applications of Peer Review Within OST, 91
OST's "Organizational" Culture and Leadership, 92

REFERENCES   94

APPENDIXES

A   Description of OST's Technology Investment Decision Model   97

B   Description of OST's Review Program   103

C   Biographical Sketches   109

D   Acronyms   113

Executive Summary

The Office of Science and Technology (OST) of the U.S. Department of Energy's (DOE's) Environmental Management (EM) Program promotes the development of new and improved technologies to lower cleanup costs and risks and to improve cleanup capabilities throughout the nation's nuclear weapons complex. The annual budget for technology development activities within OST in fiscal year 1998 is approximately $220 million, which supports more than 200 research and development (R&D) projects at universities, national laboratories, and private-sector companies. These projects are chosen for new and continued funding through a complex technology selection process, which uses the results from various types of reviews, including programmatic reviews, technical assessment reviews, and peer reviews.

Several National Research Council (NRC) committees evaluated DOE–OST's technology selection process and recommended that OST develop and apply an independent, external review process to all of its technology development programs (NRC, 1995b,c, 1996). These findings were echoed in a subsequent General Accounting Office (GAO) report, which concluded that "although the lead sites used significantly different systems to select projects, none of them used disinterested reviewers to determine the technical merit of the proposed work" (GAO, 1996, p. 7). In response to these NRC and GAO reports, in October 1996, OST instituted a peer review program to assess the scientific merit of its technology projects. According to OST, the peer review program "is designed to provide unimpeachable technical reviews on a timely basis to assist in decision making" (DOE, 1998b, p. 1). In establishing this peer review program, OST chose to use the American Society of Mechanical Engineers (ASME), with administrative and technical support provided by the Institute for Regulatory Science (RSI), to conduct peer reviews of technologies (or groups of technologies) at various stages of development.

OST asked the NRC to convene a committee to evaluate the effectiveness of its new peer review program and to make specific recommendations to improve the program, if appropriate. This is the second of two reports prepared by this committee. In its interim report published in October 1997 (NRC, 1997b), the committee conducted a preliminary assessment of OST's new peer review program. In this report, the committee provides a more complete assessment of the program after its first complete annual cycle.

DEFINITION OF PEER REVIEW

Peer review is used throughout the scientific and engineering communities to evaluate the technical merit of research proposals, projects, and programs. In this report the committee has used the following definition of peer review from the U.S. Nuclear Regulatory Commission (USNRC), which articulates the scientific and engineering communities' sense of the term:

> A peer review is a documented, critical review performed by peers [defined in the USNRC report as "a person having technical expertise in the subject matter to be reviewed (or a subset of the subject matter to be reviewed) to a degree at least equivalent to that needed for the original work"] who are independent of the work being reviewed. The peer's independence from the work being reviewed means that the peer, a) was not involved as a participant, supervisor, technical reviewer, or advisor in the work being reviewed, and b) to the extent practical, has sufficient freedom from funding considerations to assure the work is impartially reviewed. (USNRC, 1988, p. 2)

The term "peer review" has the following characteristics: expert (including national and international perspectives on the issue), independent, external, and technical.

BENEFITS OF PEER REVIEW

Peer review is recognized as an effective tool that R&D program managers can use to obtain high-quality technical input to decisions on allocating their resources (NRC, 1995a; Committee on Economic Development, 1998).

This is especially important in situations of constrained funding, where program managers are required to make decisions on the relative merit of projects within their program's R&D portfolio. If its results are used as a significant input into programmatic decision making, peer review can improve both the *technical quality* of projects in a R&D program and the *credibility* of the decision-making process. In the case of OST, such improvements may increase the likelihood that the program will produce technologies that prove effective in cleaning up contaminated sites throughout the nation's nuclear weapons complex. The 1995 report by the NRC's Committee on Environmental Management Technologies recommended the development and implementation of such a peer review program for OST's technology development program for just this reason (NRC, 1995b).

Improving Technical Quality

The independence of peer reviewers makes them more effective than internal reviewers because experts who are newly exposed to a project often can recognize technical strengths and weaknesses, and can suggest ways to improve the project that may have been overlooked by those close to it (Bozeman, 1993). Peer review can improve the technical quality of projects in a R&D program in two ways: (1) by identifying projects that lack technical merit (or are technically inferior to other feasible alternatives) so that they can be discontinued early in the R&D cycle (before large investments of funds are made), and (2) by identifying specific ways to improve proposed or ongoing projects. As a result, a greater number of alternative projects can be supported in the early stages of the development cycle, thus increasing options and chances of ultimate success in meeting the program's objectives.

Improving the Credibility of the Decision-Making Process

When peer review results are used to improve the quality of a decision process (e.g., selection of proposals, prioritization of projects for funding), they also enhance the *credibility* of the decisions. External experts often can be more open, frank, and challenging to the status quo than internal reviewers, who may feel constrained by organizational concerns. Evaluation by external reviewers thus can enhance the credibility of the review process by avoiding both the reality and the appearance of conflict of interest (Kostoff, 1997a). In addition, peer reviews that are conducted publicly, using known reviewers and following an established process that provides immediate feedback in the reviewers' own words, can enhance credibility by increasing confidence in the review process (NRC, 1997a; Royal Society, 1995).

For all of these reasons, the use of peer review increases the likelihood that decisions are consistent with the best available scientific and technical information. Of course, peer reviews in and of themselves cannot ensure the success of a project or program. Effective peer review, however, can increase the probability of project and program success. Realization of the benefits of peer review requires that the process of peer review be effective and credible and that its results be used as important input in making decisions regarding future support for the reviewed project (Chubin and Hackett, 1990).

PEER REVIEW PROCESS

The peer review process can be broken down into five general steps. For a peer review process to be credible and effective as a whole, each of these steps should be performed following well-defined procedures that are understood and accepted by those involved with the process.

1. *Selection of proposals, projects, or programs to be reviewed.* In cases where funding limitations or other factors do not allow all projects (or proposals) to be reviewed regularly, peer review program managers must have a systematic and credible approach for selecting which projects are peer reviewed. An effective selection process employs well-defined criteria to prioritize those activities to be peer reviewed.

2. *Definition of objectives of the peer review and selection of specific review criteria.* The goals, or objectives, of the peer review also must be spelled out clearly so that they are understood by those involved in the process (Chubin, 1994; Chubin and Hackett, 1990; Kostoff, 1997b). For peer reviews of projects, the objectives and utility of peer review vary with the stage of the technology development, adoption, and implementation processes. Although peer review is especially useful at the outset of a project, it can play an important role even at later stages of development and in the implementation phase, where the objectives of the peer review might be to enable late-stage refinements of the technology or to validate expectations of performance. The specific review criteria should be defined prior to the selection of peer reviewers to ensure that reviewers as a whole have the appropriate expertise. Because peer reviews are by definition technical in nature, both the objectives of the review and the review criteria should focus on technical considerations.

3. *Selection of the peer review panel.* The process for selecting reviewers must consider the fundamental characteristics of peer review and the specific objectives and criteria for the particular review being organized, and should be conducted by a person or group independent of the group being reviewed[1] (Cozzens, 1987; Koning, 1990). Peer reviewers should be selected in accordance with formally established qualification criteria that include at a minimum the following: relevant demonstrated experience, peer recognition, knowledge of the state of the art of the subject matter under review, absence of a real or perceived conflict of interest, and bias[2] such that the panel as a whole is balanced.

4. *Preparing and conducting the peer review.* For the peer review to be objective and effective, reviewers should receive written documentation that describes the proposed project and its significance and a focused charge that describes the purpose of the peer review and the review criteria. These materials should be provided to the reviewers well in advance of the review. In cases where a review panel is convened, the panel should be provided with clear presentations by the project team, as well as adequate time to assess the project comprehensively so that the panel is able to write a report that effectively summarizes and supports its conclusions and recommendations. Complete respect for confidentiality of proprietary information during review is vital. Confidentiality issues can be dealt with through panel selection (i.e., by avoiding reviewers with conflicts of interest) and by requiring panel members to formally agree not to use any such information without written permission from the author or proposer.

5. *Use of peer review results in decision making.* A peer review program will be effective only if its results are an important factor in making program decisions (Bozeman, 1993; Cozzens, 1987). Peer review reports that clearly provide the rationale for their conclusions and recommendations are an essential first step in achieving this objective. If a peer review has been planned for use in decision making, as recommended in this report, the use should be straightforward.

---

[1] Reviewers should not be selected by persons connected to the projects being reviewed (e.g., principal investigators, project managers). In cases where program managers are experts in the subject matter of the peer review and are not involved in the projects themselves (e.g., at the National Science Foundation or National Institutes of Health), however, they can be involved in the reviewer selection process.

[2] "Bias" refers to an inclination of one's outlook or point of view due to the nature of one's background, experience, and connections.

USES OF PEER REVIEW

Peer review can be employed for a variety of purposes. The following examples are illustrative, not prescriptive, and could be applied to many types of technology development programs in addition to OST's:

1. proposal evaluations—peer reviews of proposed R&D projects;
2. project maturity evaluations—peer reviews conducted as a project develops from a research idea to a technology that can be demonstrated, for example, at the following stages:

   - entrance into applied research,
   - entrance into engineering development,
   - entrance into demonstration, and
   - predeployment;

3. program balance evaluations—peer reviews that assess whether the technology development program adequately addresses the technology needs, given the resources available in the context of other competing programs; and
4. "needs" determinations—peer reviews of technology development needs; in OST's case, a review of R&D needs to address environmental problems at contaminated sites in the context of generally available technologies in the public and private sectors (both national and international).

These potential applications of peer review, along with examples of peer review programs in select organizations, are described more fully in Chapter 3.

ASSESSMENT OF OST'S PEER REVIEW PROGRAM

The committee finds that OST has made significant improvements in its peer review process since the program began in October 1996. Throughout the committee's study, OST has continued to change its peer review procedures in an effort to improve the program's effectiveness. In particular, OST has revised its review criteria to focus on technical issues, has developed a more systematic approach for selecting projects to be reviewed, and has modified its list of required documentation for the peer reviews. OST also has made a number of policy changes since this committee issued its interim report in October 1997 (see Table 1 and the main body of this report for more details on these policy changes). Although in many cases it is too early to judge the actual results of these changes, the committee is encouraged that OST senior management appears to be committed to this improvement process.

Linkage of Peer Reviews to Management Decisions

Despite the marked improvements in the *procedures* for conducting peer reviews over the past year, OST's peer review program still has not fully achieved its stated objectives of providing high-quality technical input *to assist in decision making*. The committee has found that in many cases the results of the peer reviews have not been used as input into program management and decision making. Peer reviews have been conducted on projects after OST already had decided and committed to fund the project's next stage of development, or even after a project was virtually completed. In other cases, projects on which millions of dollars had already been spent to construct pilot plants were peer reviewed before the pilot plants collected any data, that is, at a point in the project's development when no decision was to be made. In sum, peer review has in at least some instances *not* been used as an OST management tool, which is the stated objective of the peer review program.

The committee made several recommendations to address this issue in its interim report, and OST has made a number of policy changes in response (see Table 1). Although it is still too early to evaluate the impact of these policy changes, it is encouraging that OST leadership has responded to the committee's interim report by taking such actions. The linkage between peer review results and OST's decision-making process also could be improved by explicitly identifying where and how the results of peer reviews will be used, *before* the review is conducted. Therefore, **the committee recommends that as part of the documentation provided to peer review program management during the process of selecting projects for review, OST program managers be required to clearly identify the upcoming decision or milestone for which the results of the peer review will be used.** This information also should be provided to peer reviewers as part of the documentation that they receive in preparation for the review.

Selection of Projects for Review

OST has conducted peer reviews on only a small percentage of the projects currently funded within its program. As of May 1, 1998, 43 out of 226 active projects had been peer reviewed. As a result, OST developed three specific criteria to select candidate technologies for the peer review program in planning for fiscal year 1998 (FY98) peer reviews: (1) projects that were at Gate 4 or higher, (2) projects that have been supported for more than three years without being peer reviewed, or (3) projects that were new starts in FY97 or FY98. After applying these criteria, OST's peer review program staff ranked the projects in each focus area/crosscutting area (FA/CC)[3] based on the amount of funding received by each project. These ranked lists were then used by the FA/CC program managers to determine which projects were to be peer reviewed in FY98.

OST's approach addresses two primary issues identified by the committee in its interim report: the need to focus short-term peer review efforts on high-budget, late-stage projects that have never been peer reviewed, and the need to review all proposals for new technologies as they enter the project development cycle. Although OST's three selection criteria are reasonable and should help OST choose projects to be reviewed, however, they do not explicitly address two issues that have been emphasized recently by OST management: (a) the deployment of new technologies in the field, and (b) the need to reduce funding levels due to budget cuts. **To address these issues, the committee recommends that OST adopt two additional criteria to choose from those projects that satisfy one of the three existing selection criteria: (1) technologies that are being considered for deployment, and (2) technologies for which a request for further funding has been received or is anticipated.**

Although the two additional selection criteria recommended by the committee would assist OST in identifying those projects for which peer review is of highest priority, application of these criteria would still leave a large number of projects that are not peer reviewed. **To address this issue, the committee recommends that OST expand its practice of evaluating a number of related technologies in a single peer review, whenever possible.** This issue is addressed more fully in Chapter 6 (see also "Reducing the Backlog," below).

---

[3] OST's FA/CCs are administrative units used by OST to manage and coordinate its technology development activities. The four focus areas (based on OST's major problems) are: (1) mixed waste characterization, treatment, and disposal; (2) radioactive tank waste; (3) subsurface contamination; and (4) decontamination and decommissioning. The five crosscutting and supporting technology areas (technologies that apply to multiple focus areas) are: (1) robotics; (2) efficient separations; (3) characterization, sensors, and monitors; (4) industry and university programs; and (5) technology integration.

Review Criteria

A successful peer review program requires well-defined criteria to evaluate the project or program being reviewed. In response to this committee's interim report, in February 1998 OST issued a list of four "core technical criteria" for the peer review program: (1) relevance, (2) scientific and technical validity, (3) nonduplicative or superior to alternatives, and (4) data validity (see Chapter 4 for details of the criteria). The new policy states that these core technical criteria must be augmented and particularized by technology-specific criteria, to be developed by FA/CC program managers and approved by the ASME Peer Review Committee (DOE, 1998b). **The committee finds that these revised general criteria and the procedure for developing technology-specific criteria are a meaningful improvement over the original review criteria because they allow OST's peer review program staff to focus the reviews on important technical issues. The procedure also has sufficient flexibility to allow the review criteria to vary as a function of the stage of development of a technology** (see Chapter 5).

Selection of Reviewers

The selection of reviewers is one of the most important steps in the peer review process. In assessing an individual's qualifications for participation as a peer reviewer, all relevant career experience, published papers, patents, and participation in professional activities should be considered. It is also important to consider the individual's experience with peer review itself, especially for selection of the chair of a peer review panel. The group of peer reviewers should be balanced by including individuals with an appropriate range of knowledge and experience to provide credible and effective peer review of the technologies being judged. This is particularly important for the diverse and complex technologies being developed for environmental cleanup of the DOE complex. Although the range of expertise on peer review panels observed by this committee has been acceptable, the committee believes that the current databases used to identify potential reviewers (see Chapter 4, "Selection of Reviewers") may not have adequate scope to identify the broad range of reviewers likely to be necessary for some complex reviews. **The committee therefore recommends that OST establish a more systematic approach for accessing reviewer information from other databases (e.g., chemical engineers, geologists, physicists, materials scientists, biologists) or from other professional** societies, as needed to ensure the appropriate range of expertise for all review panels.

In the *ideal* situation, peer reviewers, being fully independent and external, should have no conflicts of interest. That is, they should have no current or previous relationships with the principal investigators, their organization, their proposed project, or competing projects or technologies that would impair their ability to provide an objective review. However, for various real-world reasons — for example, because contractors have many divisions and technical professionals change jobs — there can be at least appearances of conflicts. OST recently revised its conflict-of-interest policy to explicitly exclude "all DOE staff and contractors with real or potential conflicts" from consideration as peer reviewers (DOE, 1998, p. 7); *in practice*, OST has interpreted this policy to exclude all DOE staff, whether or not there is a real or potential conflict of interest with the projects under review. This interpretation goes beyond the committee's recommendation in its interim report (see Table 1) but is consistent with the ASME conflict-of-interest policy. The committee would like to point out that concerns over conflicts of interest should not necessarily preclude all DOE staff and contractors from serving on a peer review panel, however. The committee believes that DOE staff from organizations outside OST and its contractors (e.g., staff at DOE national laboratories) could be used in special circumstances when the appropriate expertise is not available outside DOE and when these individuals have had no connection with the projects under review. The reviewer selection process should in general avoid DOE staff as peer reviewers, however, and should ensure that the DOE-affiliated persons are never more than a small fraction of a panel's membership.

Documentation for Peer Reviews

The documentation required for an OST peer review is listed in Chapter 4. This list identifies some of the documents required to evaluate the technical merit of a project and, if implemented, should improve the quality of background materials provided to peer review panels. One document that is not included in this revised list of required materials is a statement of work, or proposal, describing the specific activities that will be carried out if the project is funded. **The committee recommends that a detailed proposal or statement of work be required for all peer reviews.**

Openness of Peer Reviews

The committee also encourages OST to continue to promote openness of its peer reviews and to fully inform the public and others attending the reviews of their nature. The committee believes that the strengths of open reviews (e.g., enhanced credibility of the process, and the potential for more constructive evaluations) far outweigh the potential weaknesses (e.g., possible lack of candor by some reviewers when evaluating weak proposals), especially for the peer review of projects or programs (see "Anonymous Versus Open Peer Reviews," Chapter 2).

REDUCING THE "BACKLOG"

As noted previously, OST had peer reviewed only 43 out of 226 projects that were receiving funding from the program as of May 1, 1998. Thus, there is a large backlog of projects that have never been peer reviewed. OST's current practice, in which nearly all peer reviews include formal presentations by the project team, followed by deliberations by the panel, and further question-and-answer sessions over the course of two to three days, places significant limits on the number of projects that can be peer reviewed. Even if the number of projects that were peer reviewed during a single review could be increased by improved efficiency, OST's backlog of technologies that have never been peer reviewed still would take years to address through the current process. If OST is to fulfill its policy that "all projects are to be peer reviewed" in the short term (i.e., the next year), it will have to make significant changes in how peer reviews are conducted.

The committee recommends that OST consider adopting a "triage" approach that would allow far greater numbers of technologies to be peer reviewed. This approach would involve a formal prescreening of projects by peer reviewers based exclusively on the written documentation on the project — in effect, a "mail review" of projects, followed by a formal meeting of the panel to discuss and rank them. During the prescreening review, panel members would be asked to rank all related technologies in a given area that are being considered for additional development or deployment.[4] Rankings from the panel as a whole could then be used by OST program managers to determine those highly ranked, low-budget projects that should be *considered* for funding without additional

---

[4] If the prescreening review involves technologies at very different stages of development, it might be necessary to use somewhat different review criteria for each general stage of development. The same reviewers could perform all of the reviews, however.

peer review; those highly ranked projects that should receive a more detailed evaluation (including presentations by the project team and question-and-answer sessions); and those technically weak projects that should not be considered for funding. This approach would provide OST program managers the basis for discontinuing funding for technically weak projects, and might provide them with sufficient technical input (to be supplemented by input on nontechnical factors) to make a decision to fund a low-budget project. The prescreening evaluations should not be used as the sole means for providing technical input into decisions to fund high-budget, environmental remediation projects, however. Peer reviews involving presentations by the project team and question-and-answer sessions should be carried out for all projects involving significant capital investment by OST.

Because prescreening evaluations require only written documentation on the projects to be reviewed, the triage approach could be used to evaluate a large number of projects at a single review. For example, a single prescreening review could evaluate all projects developed to address a specific type of environmental management problem or an entire OST FA/CC program. The approach also could include related projects from OST's inventory of nearly 600 projects that are not currently funded. This might allow OST to identify especially promising technologies within its inventory that should be funded for demonstration or deployment.

EVALUATION AND IMPROVEMENT MECHANISMS

In order to continue to develop and achieve a more effective peer review program, OST leadership will have to commit to a process of continuous assessment and improvement involving cycles of planning, execution, and evaluation. One approach for guiding the development of an internal evaluation procedure for the peer review program would be for OST peer review program managers to proactively seek out and learn from other organizations that have more mature peer review processes. This process of learning from the practices of other organizations is called "benchmarking." Peer review programs in some organizations that could be used by OST in such a benchmarking process are described in Chapter 3. In addition, metrics can be used to assist in the measurement of effectiveness and can help evaluate the success of a program in realizing its objectives. **The committee recommends that OST management develop an effective evaluation and improvement process for the peer review program that includes regular benchmarking against other peer review** programs and the collection of activity and performance metrics (benchmarking and metrics are discussed more fully in Chapter 7).

POTENTIAL APPLICATIONS OF PEER REVIEW WITHIN OST

Because OST has chosen to focus its new peer review program on the review of individual projects at various stages of development, the committee also has focused its reports on the peer review of projects. However, the fundamental principles of peer review (i.e., independent, external, expert, technical) also can be applied to programs or subsets of programs rather than individual projects. One potential application of such a peer review would be to evaluate R&D efforts that are needed to address the environmental problems at contaminated sites in the context of technologies available internationally in both the private and the public sectors. Another potential application would be to assess the technical balance of OST programs in the context of other programs, both within DOE (and OST in particular) and outside the DOE complex. Chapter 3 includes an overview of special characteristics of these types of reviews, including the types of reviewers and review criteria required for credible peer reviews. Although the committee discusses these issues at some length in this report, additional peer reviews should not be added to the large number of reviews currently used within OST without first evaluating the objectives and effectiveness of existing reviews. **The committee recommends that OST carefully evaluate the objectives and roles of all of its existing reviews, and then determine the most effective use of peer reviews (of various types) in meeting its overall objectives.**

OST'S "ORGANIZATIONAL" CULTURE AND LEADERSHIP

Improving the linkage of peer review results to OST's decision-making process will require more than these and other procedural changes recommended in this report. Realization of the potential benefits of peer review to the technology development program also will require a "culture change" within OST, whereby staff at all levels understand the potential benefits of peer review and incorporate peer review as an essential part of the decision-making process. Individual members of the OST organization will value peer review when they see beneficial results (e.g., which might be disseminated by using case histories), when management gives them logical messages of the value of peer review, and/or when they have incentives to use it or disincentives not to use it (Kostoff,

1997b). The committee recommends that OST leadership develop an explicit strategy to accomplish a change in its organizational culture by distributing (1) educational materials that summarize the basic principles and benefits of peer review as a tool for decision making, (2) case histories illustrating how peer review input served to improve specific projects, and (3) summaries of key performance metrics that demonstrate how peer reviews are used to meet the overall objectives of OST's program.

TABLE 1 Recommendations from Interim Report, OST's Response, and New Findings and Recommendations

Interim Report Recommendation(s)	OST Response(s)	New Finding(s) and Recommendation(s)
Overall Assessment OST has made progress in its implementation of the peer review program. To fully achieve the objectives of the program, however, OST must continue to address a number of key issues that hinder the program's successful implementation.	OST has revised its review criteria to focus on technical issues, has developed a more systematic approach for selecting projects to be reviewed, and has modified its list of required documentation for peer reviews (see details below). OST also has made a number of policy changes (see details below). Although in many cases it is too early to judge the actual results of these changes, the committee is encouraged that OST senior management appears to be committed to this improvement process.	OST has made significant improvements in its peer review process since the program began in October 1996. Despite the marked improvements in the procedures for conducting peer reviews over the past year, OST's peer review program still has not fully achieved its stated objectives of providing high-quality technical input to assist in decision making.
Linkage to Decision Making The committee . . . encourages OST to enforce the 30-day requirement for written responses, and to require more detailed responses that fully describe how the recommendations of the peer review reports were implemented or considered.	The EM-53 Office Director now coordinates with program managers regarding their written responses to peer review reports. Detailed responses to the review panel's recommendations are now required.	The committee recommends that as part of the documentation provided to peer review program management during the process of selecting projects for review, OST program managers be required to clearly identify the upcoming decision or milestone for which the results of the peer review will be used.

TABLE 1 (continued)

Interim Report Recommendation(s)	OST Response(s)	New Finding(s) and Recommendation(s)
Application of Peer Review OST should restrict the term "peer review" to only those technical reviews conducted by independent, external experts. OST should adopt alternative terms, such as "technical review," for its internal reviews of scientific merit and pertinency.	OST's Implementation Guidance states that OST is restricting the term peer review to only those technical reviews conducted by independent, external experts.	Although the committee discusses a number of potential applications of peer review, additional peer reviews should not be added to the large number of reviews currently used within OST without first evaluating the objectives and effectiveness of existing reviews. The committee recommends that OST carefully evaluate the objectives and roles of all of its existing reviews, and then determine the most effective use of peer reviews (of various types) in meeting its overall objectives.
Selection of Projects to be Reviewed OST should develop a rigorous process for selecting projects to be peer reviewed. To be fair and credible, this process should employ well-defined project selection criteria, and OST peer review program staff should be directly involved in making decisions regarding which projects will be reviewed.	The Peer Review Coordinator is to recommend a rigorous process for selecting projects to be peer reviewed. In FY98, OST developed three specific criteria to select candidate projects for peer review: (1) projects that were at Gate 4 or higher, (2) projects that had been funded for more than three years, or (3) projects that were new starts in FY97 or FY98. Program managers then prioritized projects for peer review from among the list of candidate projects.	The three selection criteria are reasonable and should help OST choose projects to be reviewed; however, they do not explicitly address OST's goal of deploying new technologies or the need to reduce funding levels due to budget cuts. The committee recommends that OST apply two additional criteria to choose from among those projects that satisfy one of the three existing selection criteria: (1) technologies that are being considered for deployment, and (2) technologies for which a request for further funding has been received or is anticipated. The committee also recommends that OST expand its practice of evaluating a number of related technologies in a single peer review, whenever possible.

Review Criteria The committee... recommends that OST revise... general nontechnical criteria to focus on technical aspects of these issues, or to remove them from the list of review criteria. The committee recommends that OST develop a well-defined general set of technical criteria for peer reviews, to be augmented by technology-specific criteria as needed for particular reviews.	OST generated a new list of general review criteria that do not include nontechnical issues. Core criteria must be supplemented by technology-specific criteria developed by the FA/CC program manager requesting the review.	The committee finds that OST's revised general criteria and the procedure for developing technology-specific criteria are a meaningful improvement over the original review criteria because they allow OST peer review program staff to focus the reviews on important technical issues. The procedure also has sufficient flexibility to allow the review criteria to vary as a function of the stage of development of a technology.
Objectives of Reviews The committee encourages OST to use the statements of desired "scope" and "purpose" that are developed by the FA/CC program managers to communicate the objectives of the review to all involved in the peer review. The committee also recommends that the objectives of the review be clearly stated in the review report.	OST's new policy states that the objectives of the review (or "Charter of the Review Panel") are to be included in every peer review report, although none of the reports issued through April 1998 included such a description.	The committee concludes that OST has developed an appropriate policy but has not yet implemented it, and therefore encourages OST to follow up on this new requirement.
Selection of Reviewers OST should consider modifying reviewer selection criteria to emphasize expertise relevant to the review. In addition, the size of the review panel should depend on a number of factors, including the complexity and number of projects being reviewed and the specific review criteria established for the peer review. One approach that could be used to help identify reviewers with relevant expertise would be to develop and use a large data base of potential reviewers.	RSI uses several sources of names of potential reviewers, including a data base of past reviewers, reviewer nominees, an RSI database of contacts, and the ASME membership list.	The committee recommends that OST establish a more systematic approach for accessing reviewer information from other databases (e.g., chemical engineers, geologists, physicists, materials scientists, biologists) and other professional societies, as needed to ensure the appropriate range of expertise for all review panels.

TABLE 1 (continued)

Interim Report Recommendation(s)	OST Response(s)	New Finding(s) and Recommendation(s)
Conflict of Interest In order to ensure the independence of the peer reviewers . . . the committee recommends that OST also include a criterion that explicitly excludes EM staff and contractors with real or potential conflicts of interest, including all OST staff and contractors, from consideration as peer reviewers.	All DOE staff and contractors with real or potential conflicts of interest are explicitly excluded from consideration as reviewers. In practice, OST has interpreted this policy to exclude all DOE staff, whether or not there is a real or potential conflict of interest.	This restrictive standard goes beyond the committee's recommendation but is consistent with the ASME Peer Review Committee's interpretation of ASME's conflict-of-interest policy. Concerns over conflicts of interest should not necessarily preclude all DOE staff and contractors from serving on a peer review panel, however. The committee believes that DOE staff from organizations outside OST and its contractors (e.g., staff at DOE national laboratories) could be used in special circumstances when the appropriate expertise is not available outside DOE and when these individuals have had no connection with projects under review.
Documentation for Peer Reviews The committee encourages OST to refine its list of required materials to include only technical documents relevant to the peer review criteria established for the peer review.	OST has modified the required documentation to fit with the new general review criteria and has removed nontechnical documents from the list.	One document that is not included in the revised list of required materials is a statement of work, or proposal, describing the specific activities that will be carried out if the project is funded. The committee recommends that a detailed proposal or statement of work be required for all peer reviews.
Peer Review Reports The committee believes that the peer review reports could be improved further . . . by also including a statement of the objective of the review and a list of references used in the analysis.	OST policy states that the format of the review report will be changed to include these items, although the committee has not reviewed a technical report prepared completely in the new format.	The committee concludes that OST has developed an appropriate policy but has not yet implemented it, and therefore encourages OST to follow up on this new requirement.

Reducing the Backlog
Because the majority of projects are in the late stages of development (past Gate 4), OST should concentrate reviews on high-budget, late stage projects that have never been reviewed, but that still face upcoming decisions with major programmatic and/or funding issues.

Management of Peer Review Program
The committee recommends that OST develop a targeted plan for the peer review program. Such a plan should consider factors such as how many of OST's technology projects can be peer reviewed, realistic schedules for the reviews, and the peer review program budget. To be effective, this plan also must assure that peer reviews are conducted early enough in the budget cycle to allow peer review results to be used as an input into meaningful funding decisions.

To develop its FY98 schedule from a list of 182 projects that met one of its three new selection criteria, OST sorted the projects by amount of funding and then provided lists to FA/CC managers who were asked to prioritize the lists by identifying those projects for which a peer review would be most valuable for decision making.

The Peer Review Coordinator has been tasked with developing a targeted plan for the peer review program. The Peer Review Coordinator has developed a draft schedule for FY98 peer reviews.
The EM-53 Office Director now coordinates with the program managers regarding their written responses to peer review reports. Detailed responses to the review panel's recommendations are now required.

The committee recommends that OST consider adopting a triage approach that would allow far greater numbers of technologies to be peer reviewed.

The committee recommends that OST management develop an effective evaluation and improvement process for the peer review program that includes regular benchmarking against other peer review programs and the collection of activity and performance metrics.

TABLE 1 (continued)

Interim Report Recommendation(s)	OST Response(s)	New Finding(s) and Recommendation(s)
Organizational Culture and Leadership The peer review program will be meaningful only if OST acknowledges the benefits of effective peer review and makes peer review a vital part of the decision and management process throughout the organization. Although attaining these benefits will require a sustained effort from management, the entire organization will be rewarded through enhancement of the technology-development program.	OST has improved its documentation of the peer review program, including its "Implementation Guidance for the Office of Science and Technology Technical Peer Review Process" document and the "The Office of Science and Technology Peer Review Process: At a Glance" flyer.	The committee recommends that OST leadership develop an explicit strategy to accomplish a change in its organizational culture by distributing (1) educational materials that summarize the basic principles and benefits of peer review as a tool for decision making, (2) case histories illustrating how peer review input served to improve specific projects, and (3) summaries of key performance metrics that demonstrate how peer reviews are used to meet the overall objectives of OST's program.

1

Introduction

The U.S. Department of Energy's (DOE's) Environmental Management (EM) Program was established in 1989 to address the health, safety, environmental, and regulatory challenges associated with cleanup of the nation's nuclear weapons complex. Within EM, the Office of Science and Technology (OST) was created to promote the development of improved technologies to lower cleanup costs and risks (to workers, the public, and the environment) and to improve cleanup capabilities. OST supports the entire range of technology development activities—beginning with basic research through the EM Science Program (NRC, 1997a) and extending through development, demonstration, and (with the assistance of industrial partners) deployment into the cleanup program.

In fiscal year 1998 the annual budget for EM is about $5.6 billion, of which about $220 million is devoted to technology development activities within OST. The importance of technology development to EM's mission has been recognized in *Accelerating Cleanup: Paths to Closure* (DOE, 1998a), which describes EM's plan for cleanup of the weapons complex. This plan discusses the importance of technology development to reduce the "mortgage" at the complex—the long-term costs of maintaining contaminated buildings, equipment, and sites until they can be remediated.

OST sponsors 226 active research and development projects at universities, national laboratories, and private-sector companies on topics ranging from the remote detection of contaminants in the subsurface using geophysical techniques to the development of melters for waste vitrification. The major research and development program units within OST are listed in Table 1.1. OST uses various types of reviews (e.g., programmatic reviews, technical assessment reviews, peer reviews[1]) in its technology selection process

---

[1] These peer reviews are termed "technical peer reviews" by OST.

(see Appendix B for a description of the different types of reviews). These reviews are used to assess the merit of individual projects as well as the merit of all technology development thrusts within the office.

TABLE 1.1: DOE–OST R&D Program Units

Office of Technology Systems[a]	Office of Science and Risk Policy
Decontamination and decommissioning focus area	Environmental management science program (EMSP)
Mixed waste characterization focus area	Risk policy program
Radioactive tank waste focus area	
Subsurface contamination focus area	
Characterization, monitoring, and sensors cross cut	
Efficient separations program cross cut	
Robotics cross cut	
Industry and university programs	
Technology integration	

[a] OST's focus areas, cross cuts, and supporting technology areas are administrative units used by OST to manage and coordinate its technology development activities, and are based on DOE–EM's major problems.

OST'S PEER REVIEW PROGRAM

Several recent National Research Council (NRC) reports evaluated OST's technology selection process and recommended that OST develop an independent, external review process and apply it to all technology development programs. The report *Improving the Environment* recommended that "technology selection should incorporate a knowledgeable independent review group that has no vested interests in the outcome and that includes people from outside the Department who work in the commercial use of technologies" (NRC, 1995c, p. 104). The NRC's Committee on Environmental Management Technologies (CEMT) also evaluated this technology selection process in its 1994 and 1995 annual reports (NRC, 1995b, 1996). In particular, these reports

INTRODUCTION

recommended that OST develop a standardized, rigorous, and independent external peer review process and apply it to all technology development programs. These findings were echoed in a subsequent General Accounting Office (GAO) report, which concluded that "although the lead sites used significantly different systems to select projects, none of them used disinterested reviewers to determine the technical merit of the proposed work" (GAO, 1996, p. 7).

In response to these NRC and GAO reports, OST recently instituted a peer review program that uses the American Society of Mechanical Engineers (ASME), with administrative and technical support provided by the Institute for Regulatory Science (RSI), to conduct peer reviews of technologies (or groups of technologies) at various stages of development. According to the OST, the objective of this new program is to serve as a "management tool for assuring that the technology is of high quality and effective, that critical needs have not been overlooked, and that the technology has the best chance possible for implementation."[2]

CHARGE TO THE COMMITTEE

OST asked the NRC to convene an expert committee to evaluate the effectiveness of its new peer review program and to make specific recommendations to improve the program, if appropriate. In particular, the committee was asked to review the following:

- internal procedures used by OST to identify the need for timely peer review of projects and programs;
- structures, protocols, and procedures for obtaining peer reviews of OST projects and programs, including who decides what will be peer reviewed, what criteria for peer review are used, and when in the R&D process peer review is requested; and
- feedback of peer review results into program management and development decisions.

In performing this assessment, the committee was asked to compare OST's practices to generally accepted norms for scientific and technical peer review, including practices for selection of peer reviewers and screening for bias and conflict of interest. The assessment has been made more challenging by the

---

[2]Presentation to committee by Anibal Taboas, DOE, Washington, D.C., February 24, 1997.

fact that OST has continued to improve its peer review process throughout this study, based on input from the committee's interim report, advice from the ASME Peer Review Committee, and OST's peer review program staff.

OST has chosen to establish a relationship with two outside organizations, ASME and RSI, to carry out parts of its peer review process. In performing this assessment, the committee has focused on the structures, protocols, and procedures of OST's peer review program, including those parts carried out by ASME and RSI, but has avoided an explicit evaluation of the performance of these organizations because they are acting simply as agents of OST. The responsibility for establishing and executing an effective peer review program lies entirely with OST. The committee also has not evaluated, or endorsed, the organizational arrangements among ASME, RSI, and OST. The committee's findings and recommendations are directed at OST, which as noted above, has ultimate responsibility for all aspects of the peer review program.

PARALLEL EFFORTS

Another NRC Committee, the Committee on Prioritization and Decision Making in the DOE–OST, is currently conducting a parallel evaluation of the decision-making processes throughout OST's technology development program. One aspect of its work will be to examine the role and importance of peer reviews (and other types of reviews) in OST's decision-making processes. Our committee therefore has focused its work on OST's peer review program itself, including an evaluation of how peer reviews, if conducted, could be made more useful as an *input* to OST's decision-making processes, but has not evaluated OST's decision-making processes explicitly.

STUDY PROCESS

The committee was briefed on the newly instituted peer review program by DOE staff at six committee meetings and has reviewed the recently developed *Implementation Guidance for the Office of Science and Technology Technical Peer Review Process* (DOE, 1998b), which documents OST's revised peer review procedures. Members and staff of the committee also observed peer reviews conducted on the In Situ Redox Manipulation project in Richland, Washington; the Decontamination and Decommissioning (D&D) Large Scale Demonstration projects and three D&D Technology Development projects in

Morgantown, West Virginia; the MAG*SEP technique[3] in Atlanta, Georgia; the Small In-Tank Processing Modules and Small Modular In-Can Vitrification projects in Columbia, Maryland; several High Temperature Melter and Characterization projects in Idaho Falls, Idaho; and the Vortec Combustion Melter, Catalytic Chemical Oxidation, and Steam Reforming of Low-Level Mixed Waste projects in Columbia, Maryland. Committee members and staff also attended the Annual Meeting of the ASME Peer Review Committee in November 1997 and the ASME Peer Review Committee meeting in January 1998. In addition, the committee reviewed all peer review reports produced under the program from its initiation in October 1996 through April 1998.

In its work, the committee reviewed the literature on peer review and placed an emphasis on comparison with peer review in other organizations—an informal type of benchmarking. To aid in the comparison, during its meetings the committee was briefed on current peer review practices at the Department of Defense (DOD) Strategic Environmental Research and Development Program (SERDP), the National Institutes of Health (NIH), the National Science Foundation's (NSF's) Small Business Innovation Research (SBIR) program, the National Institute of Standards and Technology (NIST), the Environmental Protection Agency (EPA), EPA's Science Advisory Board, the U.S. Nuclear Regulatory Commission (USNRC), DOE's Office of Basic Energy Sciences (BES), Information Dynamics, Inc. (which conducts reviews for the National Aeronautics and Space Administration's [NASA's] Office of Life and Microgravity Sciences and Applications [OLMSA]), and the American Institute of Biological Sciences (AIBS). The committee also drew on its members' personal knowledge of these and other peer review programs.

REPORTS

This is the second of two reports prepared by the committee. In October 1997, the committee completed its interim report, which provided OST with a preliminary assessment of its new peer review program (NRC, 1997b). The interim report described the essential components of a credible peer review process and provided a preliminary assessment of OST's peer review program and the status of its implementation. In the interim report, the committee examined broad issues and tried to offer constructive recommendations to assist OST in successfully implementing this program, focusing on how OST could most effectively establish the basic elements of a sound peer review process.

---

[3]The MAG*SEP technique is a means of recovering selected radionuclides and heavy metals from water and other liquids through sorption onto specially coated particles.

OST has moved expeditiously to implement many of the recommendations from the interim report.

In this final report, the committee first reviews the current status of OST's peer review program (as of April 1998) by reevaluating the specific policy changes that have been implemented by OST in response to issues raised in the committee's interim report. The committee also addresses three specific issues that OST has identified as particularly important at this time:

1. How should OST deal with the large number of technology projects at late stages of development that have never been peer reviewed? (See Chapter 6.)

2. Can the fundamental principles of peer review outlined in the committee's interim report be applied to other types of OST reviews? (See Chapters 3 and 7.)

3. How can OST measure the effectiveness of its peer review program? (See Chapter 7.)

In addition, in Chapter 7 the committee discusses a number of approaches that OST could use to continue to improve the peer review program in the future.

2

Elements of a Credible Peer Review Program

In this chapter, the committee gives a general overview of the concept of "peer review" and highlights the basic elements of the peer review process. The committee presents a definition of peer review, discusses its potential benefits, and then summarizes the fundamental steps of the peer review process. The processes described in this chapter relate not only to OST's particular peer review program, but also to other peer review programs designed to evaluate the technical merit of technology development activities.

DEFINITION OF PEER REVIEW

Peer review is used throughout the scientific and engineering communities to evaluate the technical merit of research proposals, projects, and programs. Although one might argue legitimately that peer review[1] is the name given to any evaluation of technical[2] merit by other experts working in or close to the field in question, the scientific and engineering communities generally use the term in a narrower sense. In its interim report, the committee adopted the following definition developed by the U.S. Nuclear Regulatory Commission that articulates these communities' sense of peer review:

---

[1] The choice of the term "peer review" versus "merit review" is somewhat subjective. Because "merit review" is often used to describe evaluations that include programmatic/nontechnical aspects of projects (Royal Society, 1995), the committee has chosen to use the term "peer review" in this report.

[2] In this report, the committee uses the term "technical" to mean "relating to special and/or practical knowledge of an engineering or scientific nature."

A peer review is a documented, critical review performed by peers [defined in the USNRC report as "a person having technical expertise in the subject matter to be reviewed (or a subset of the subject matter to be reviewed) to a degree at least equivalent to that needed for the original work"] who are independent of the work being reviewed. The peer's independence from the work being reviewed means that the peer, a) was not involved as a participant, supervisor, technical reviewer, or advisor in the work being reviewed, and b) to the extent practical, has sufficient freedom from funding considerations to assure the work is impartially reviewed.

A peer review is an in-depth critique of assumptions, calculations, extrapolations, alternate interpretations, methodology, and acceptance criteria employed, and of conclusions drawn in the original work. Peer reviews confirm the *adequacy* of the work. (USNRC, 1988, p. 2)

In this definition, the term peer review has the following characteristics:

- expert (including national and international perspectives on the issue),
- independent,
- external, and
- technical.

Most importantly, peer reviews must be carried out by independent evaluators who are experts in the technical issues relevant to the projects under review. Such reviewers must be highly qualified[3] and independent in order to evaluate credibly the scientific and engineering merit of the project (or subset of project components) with respect to current technologies, both domestic and international. In the report *Allocating Federal Funds for Science and Technology* (NRC, 1995a, p. 69), peers are defined as "established working scientists or engineers from diverse research institutions who are deeply knowledgeable about the field of study and who provide disinterested technical judgments as to the competence of the researchers, the scientific significance of the proposed work, the soundness of the research plan, and the likelihood of success." Note that such reviewers are not necessarily expert in, or familiar with,

---

[3]Determined by reputation and standing in the field (e.g., patents, publications, and status in professional societies) and relevance to the project being reviewed.

the agency program or relevant contextual factors. These considerations are the proper province of agency management. It is important to note that internal reviews, although useful for program management, should not be confused with peer review. The independence of peer reviewers distinguishes them from internal reviewers; and thus, the term "internal peer review" is an oxymoron (Bozeman, 1993).

The USNRC's definition of "peer review" provides some guidance on the issue of potential conflict of interest by explicitly excluding potential reviewers who have been involved with the specific project being reviewed or who have financial interests in the outcome of the reviews. As Chubin and Hackett (1990) have pointed out, however, it is difficult, if not impossible, to fully separate persons with relevant expertise from those with potential conflicts of interest, because "experts" on a subject or technology almost necessarily have some interest in the outcome of the review. Dealing with such issues is a challenge to the management of any peer review program. The issue of conflict of interest is discussed more fully below in the section "Selection of Reviewers."

BENEFITS OF PEER REVIEW

Peer review is used throughout the scientific and engineering communities to evaluate the technical merit of research proposals, projects, and programs. Peer review is recognized as an effective tool that R&D program managers can use to obtain high-quality input to decisions on allocating their resources (Committee on Economic Development, 1998; NRC 1995a). This is especially important in situations of constrained funding, where program managers are required to make decisions on the relative merit of projects within their program's research and development portfolio. If its results are used as a significant input into programmatic decision making, peer review can improve both the *technical quality* of projects in a research and development program and the *credibility* of the decision-making process. In the case of OST, such improvements may increase the likelihood of the program's producing technologies that prove effective in cleaning up contaminated sites throughout the nation's nuclear weapons complex. The 1995 NRC CEMT report recommended development and implementation of such a peer review program for the OST technology development program for just this reason (NRC, 1996).

Improving Technical Quality

The independence of peer reviewers makes them more effective than internal reviewers because experts who are newly exposed to a project often can recognize technical strengths and weaknesses, and can suggest ways to improve the project that may have been overlooked by those close to it (Bozeman, 1993). Peer review can improve the technical quality of projects in a research and development program in two ways: (1) by identifying projects that lack technical merit (or are technically inferior to other feasible alternatives) so that they can be discontinued early in the R&D cycle (before large investments of funds are made), and (2) by identifying specific ways to improve proposed or ongoing projects. As a result, a greater number of alternative projects can be supported in the early stages of the development cycle, thus increasing options and chances of ultimate success in meeting the program's objectives. In studies of corporate product development programs, Cooper (1993) has shown that such early decisions can result in a greater than 50 percent overall increase in productive efficiency.

Improving the Credibility of the Decision-Making Process

When peer review results are used to improve the quality of a decision process (e.g., selection of proposals, prioritization of projects for funding), they also enhance the *credibility* of the decisions. External experts often can be more open, frank, and challenging to the status quo than internal reviewers, who may feel constrained by organizational concerns. Evaluation by external reviewers thus can enhance the credibility of the review process by avoiding both the reality and the appearance of conflict of interest (Kostoff, 1997a). In addition, peer reviews that are conducted publicly, using known reviewers and following an established process that provides immediate feedback in the reviewers' own words, can enhance credibility by increasing confidence in the review process (NRC, 1997b; Royal Society, 1995).

For all of these reasons, the use of peer review increases the likelihood that decisions are consistent with the best available scientific and technical information. Of course, peer reviews in and of themselves cannot ensure the success of a project or program. Effective peer review can increase the probability of project and program success, however. Realization of these benefits requires that the process of peer review be effective and credible and that its results be used as an important input in making decisions regarding future support for the reviewed project (Chubin and Hackett, 1990).

PEER REVIEW PROCESS

In the committee's interim report, the peer review process was defined by the following five general steps:

1. selection of proposals, projects, or programs to be reviewed;
2. definition of objectives of the peer review and selection of specific review criteria;
3. selection of the peer review panel;
4. preparing and conducting the peer review; and
5. use of peer review results in decision making.

In order for a peer review process to be credible and effective as a whole, each of these steps must be performed following well-defined procedures that are understood and accepted by everyone involved with the peer review process.

Selection of Projects to be Reviewed

In many types of peer review programs, the selection process is straightforward (e.g., all proposals are reviewed before they are funded.). In some organizations that use peer review to evaluate the technical merit of ongoing projects (including DOE–OST), however, decisions must be made regarding which specific projects are to be peer reviewed and at what stage in their funding or development path. In the case of DOE–OST, for example, a large number of technology projects have been funded for years before the new peer review program was initiated in October 1996. As of May 1, 1998, 43 of 226 active projects had been peer reviewed. Specific issues related to this "backlog" of OST projects that have never been peer reviewed are addressed in Chapter 6 of this report. In the general case, peer review program managers must have a systematic and credible approach for selecting which projects (or programs) are reviewed by the peer review program. An effective selection process employs well-defined criteria to prioritize those activities to be peer reviewed.

Definition of Review Objectives and Selection of Review Criteria

The goals, or objectives, of the peer review must be spelled out clearly so that they are understood by all involved in the process (Chubin, 1994; Chubin and Hackett, 1990; Kostoff, 1997b). In addition, the specific review criteria (i.e., specific questions or issues that reviewers are asked to address in a particular review) should be defined prior to the selection of peer reviewers to ensure that the review panel as a whole has the appropriate mix of expertise required to address these issues. Because peer reviews are by definition technical in nature, both the objectives of the review and the review criteria should focus on technical considerations. Reviewers and presenters should be informed of the objectives and review criteria well in advance of the review.

For project reviews, the objectives and utility of peer review vary according to the stage of the technology development, adoption, and implementation processes. It is especially useful at the outset of a project—that is, at the proposal stage, when peer review can help to select which of several candidate technologies to develop. It must be emphasized, however, that peer review can play an important role even at later stages of development (e.g., at the point where a technology project moves from bench to field scale). Even at the implementation phase of technology development, peer review can be used to validate expectations of performance and to enable late-term modifications that would enhance a technology's utility. In early phases, peer review can help in the selection of technologies. In later phases, it can lead to late-stage refinements in technologies or to the validation of technologies already developed.

To be effective, review criteria should be consistent with the objectives of the review and be clearly understood by all involved in the review (i.e., reviewers, principal investigators [PIs],[4] program managers). Many effective peer review programs employ a small number of general review criteria, and this is the standard operating mode for organizations that review a large number of proposals (e.g., NSF, NIH). For reviews of technology development projects at various stages of development, however, a flexible system that allows specific review criteria to be selected for individual projects can be effective in focusing reviewers on issues of particular interest for a given project or group of projects.

---

[4]PIs are the scientists and engineers responsible for research and development on a specific project.

Selection of Reviewers

The selection of reviewers is a critically important step in the peer review process. The process for selecting reviewers must reflect the fundamental characteristics of peer review described earlier in this chapter (see section entitled "Definition of Peer Review") and the specific objectives and criteria for the particular review being organized. Reviewer selection should be conducted by a group independent[5] of the one being reviewed (Cozzens, 1987; Koning, 1990).

Qualifications

Peer reviewers should be selected in accordance with formally established qualification criteria. In the committee's view, the minimum criteria for individual reviewers are relevant, demonstrated experience in the subject to be reviewed, including national and international perspectives on the issue (Bozeman, 1993; Porter and Rossini, 1985); and peer recognition (both nationally and internationally). In assessing an individual's qualifications for participation as a peer reviewer, all relevant career experience, published papers, patents, and participation in professional activities should be considered. The group of peer reviewers should be balanced by including individuals with an appropriate range of knowledge and experience to provide credible and effective peer review of the technology being judged (Porter and Rossini, 1985). It is also important to consider the individual's experience with peer review itself (Royal Society, 1995). In cases where a review panel is established, the chair should be internationally respected in the field under review and should be an experienced peer reviewer. Additionally, some representation of reviewers who have knowledge of competing or alternative technologies is desirable.

Conflict of Interest

The reviewer selection process also must ensure that reviewers do not have real or potential conflicts of interest (e.g., not selected from OST or the OST contractor community; see "Conflict of Interest," Chapter 5) with the

---

[5] Reviewers should not be selected by persons connected with the projects being reviewed (e.g., PIs, project managers). However, in cases where program managers are experts in the subject matter of the peer review and are not involved in the projects themselves (e.g., at NSF or NIH), they can be involved in the reviewer selection process.

activities under review or biases[6] so that the panel as a whole is balanced (Abrams, 1991; Cole, 1991; Moxham and Anderson, 1992). In the *ideal* situation, peer reviewers, being fully independent and external, should have no conflicts of interest. That is, they should have no current or previous relationships with the PIs, their organization, their proposed project, or competing projects or technologies that would impair their ability to provide an objective review. However, for various real-world reasons (e.g., because contractors have many divisions and technical professionals often change jobs, or due to the inherent "interest" of any expert), it may not be practical (or possible) to avoid at least the appearance of conflicts. In some cases, it might be necessary to impose a "statute of limitations" on conflicts of interest whereby prior association with an investigator or organization that ended several years before the review does not necessarily preclude a potential reviewer from serving on a review panel.

Problems that might arise involving conflicts of interest can be mitigated by requiring all reviewers (and other persons involved) to sign a statement disclaiming or disclosing any real or apparent conflict of interest. If necessary, conflicted reviewers could be disqualified. For example, such statements are standard in the DOE (BES) national laboratories and are required by Public Law 102-564. All public officials must comply with this act and so disqualify themselves from engaging in any transaction or decision (e.g., appointment making, voting on an issue, entering into a contractual agreement, negotiating to affect decisions) that may materially affect their financial interest. Likewise, all reviewers in a peer review should affirm that they have no conflict of interest, whether financial, personal or intellectual. Some institutions—for example, the NRC—conduct a dialogue among review panel members, during which all panel members disclose and discuss their potential biases and conflicts of interest.

Planning and Conducting the Review

For a peer review to be objective and effective, peer reviewers should receive written documentation that describes the project and its significance (i.e., why the project is being conducted and what it proposes to contribute) and a focused charge that describes the purpose of the peer review and the review criteria. These materials should be provided to the reviewers well in advance of the review. In cases where a review panel is convened, the panel should be

---

[6] Bias refers to an inclination of one's outlook or point of view due to the nature of one's background, experience, and connections.

provided with clear presentations by the project team, as well as adequate time to assess the project comprehensively so that the panel is able to write a report that effectively summarizes and supports its conclusions and recommendations.

Confidentiality of Technical Information

One of the most challenging issues involved in the reviews of technology development projects (especially those involving industrial partners) is the handling of proprietary information necessary to evaluate the technical merit of a project. Reviewers may need to receive privileged information during the review process to evaluate the technical merit of a project. Complete respect for confidentiality is central to the successful operation of peer review (Royal Society, 1995) because technical information may not have been patented or copyrighted at the time of review and therefore requires protection as the intellectual property of the authors or proposers. Confidentiality of proprietary information during review can be dealt with by panel selection (i.e., avoiding reviewers with conflicts of interest) and by requiring panel members to formally agree not to use any such information without written permission from the author or proposer. If the PI does not disclose proprietary information, the project will receive a poor review because some technical bases for the evaluation will be missing. To avoid such situations, an investigator's agreement to disclose information critical to a meaningful peer review (under appropriate confidentiality agreements) should be a condition of the initial project award.

Anonymous Versus "Open" Peer Reviews

One consideration when planning and conducting peer reviews is whether the evaluations should be conducted anonymously or openly (i.e., using publicly known reviewers). Most peer reviews of proposals are conducted using anonymous reviewers (e.g., see Boxes 3.1, 3.2, and 3.3). The principal strength of anonymous reviews is that they may encourage reviewers to be more candid and frank in their evaluations, because they are shielded to some degree from potential reprisals from unsuccessful proposers. In practice, however, anonymity is often difficult to ensure due to characteristic points of view or writing styles, particularly in highly specialized fields (Chubin and Hackett, 1990). The shielding of reviewers from proponents of a project through anonymity also may enhance the perceived "fairness" of the review process because it requires that the reviewers base their evaluations solely on the written documentation provided to them (which can be standardized in a relatively uniform format), rather than more "ad hoc" communications between the evaluators and proponents of different projects. This same lack of communication also can limit the effectiveness of anonymous peer reviews, however, because it severely limits the ability of reviewers to clarify questions that arise during the evaluation of written documentation.

Many peer reviews of programs and projects in progress are conducted using reviewers who are known to the proposers and other interested parties, and include open question-and-answer sessions, or "open reviews" (e.g., see Boxes 3.5, 3.6, and 3.7). It is important to note that the use of the term open review does not imply that all deliberative sessions are held in public. The ability of an evaluating body to frankly discuss the merits and weaknesses of a project and to reach consensus in a closed session is an important attribute of many open reviews. The principal strength of open reviews is that they allow for more detailed evaluations because reviewers are permitted to ask questions and request additional information to clarify issues that arise during the evaluation. The opportunity for interaction often allows evaluators to be more specific and constructive in their comments (e.g., by suggesting ways in which a project could be improved, rather than simply arriving at an overall ranking). Open reviews also can serve to increase the *credibility* of the review process because the process is more transparent to potential critics (e.g., government oversight bodies, advocates of unfunded projects), and because the reviewers are more accountable for their evaluations (Chubin and Hackett, 1990; Royal Society, 1995). The most significant potential weakness of open reviews is that reviewers may be less candid, especially in their evaluations of weak projects or proposals from well-known proponents, if they fear the possibility of reprisals.

Use of Peer Review Results in Decision Making

A peer review program will be effective only if its results are an important factor in making program decisions, for example, regarding future support for the reviewed project and/or as input to improve the technical merit of the project (Bozeman, 1993; Cozzens, 1987). Peer review reports that clearly provide the rationale for their conclusions and recommendations are an essential first step in achieving these objectives. If a peer review has been planned for use in decision making, as recommended in this report, the use should be straightforward. In addition, there should be procedures to monitor how project personnel follow through on technical recommendations of the peer review panel. A well-established peer review program also should have specific metrics to evaluate how well the peer review program achieves its objectives (see "Metrics," Chapter 7).

3

Uses of Peer Review

Although OST has chosen to apply its new peer review program almost exclusively to the evaluation of projects at various stages of development, the basic principles of peer review (i.e., expert, independent, external, technical) also can be used to evaluate the technical merit of proposals, the balance of programs, and even program "needs." The evaluation of proposals is by far the most common and well-established form of peer review. Peer review also can be used to evaluate the technical merit of individual projects, or groups of projects, at various stages of development (as is done within OST). In addition, the basic principles of peer review, when effectively focused and defined in terms of the purpose of a review, can be used to evaluate a whole program or aspects of a program (including program "needs"). This chapter addresses how the defining characteristics of peer review developed in Chapter 2 (i.e., expert, independent, external, technical) apply to various types of evaluations. As examples, the committee discusses four specific applications of peer reviews:

1. proposal evaluations;
2. project maturity evaluations:

   - entrance into applied research,
   - entrance into engineering development,
   - entrance into demonstration, and
   - predeployment;

3. program balance evaluations; and
4. "needs" determinations.

In this chapter, the committee provides a brief overview of each of these applications, discusses specific features or considerations for each review type, and describes a number of model review programs from other governmental and nongovernmental organizations (Boxes 3.1 to 3.8). It must be emphasized that these are illustrative, not prescriptive, examples; they are offered to illustrate how the principles of peer review might be applied broadly to different types of reviews, *not* to prescribe particular additional reviews for OST. Although some of the examples are framed in terms of OST's program, the types of reviews could be applied to many kinds of technology development programs. They also can be used for benchmarking OST approaches with other peer review programs (see Chapter 7).

PROPOSAL EVALUATIONS

Peer review is the most common method for evaluating the technical merit of proposed research and development projects. There are a number of well-known models for such reviews, including those of NSF and NIH (see Kostoff, 1997b; NSF, 1997; and OTA, 1991 for detailed descriptions of these, and other approaches for the review of research proposals). Because proposal evaluation often involves the review of large numbers of proposals, such evaluations typically employ a small number of fixed review criteria and generally utilize mail reviews. In most cases, review panels also meet as a group to supplement evaluations received by mail.

Organizations that support applied research proposals, such as those supported by OST, often choose to employ a two-stage selection process: (1) a peer review to assess the technical merit of the proposals, and (2) a "relevance" or "commercial viability" review to assess the potential applicability of the proposed project to program needs. Box 3.1 describes an example of such a two-stage selection process employed by NSF's Small Business Innovation Research Program. The EM Science Program is an example of an OST program that employs such a two-stage selection process (see NRC, 1997a).

Although many federal agencies such as NSF and NIH administer their own peer review processes, others choose to contract out part of the review process to other organizations with experience administering peer reviews. Such arrangements can be beneficial to federal programs that do not have sufficient staff expertise to administer effective, credible peer review programs or that choose to maintain an extra degree of independence from the peer review process. Examples of two such arrangements are described briefly in Boxes 3.2 and 3.3. OST has chosen to establish a similar arrangement with the American

BOX 3.1
Peer Review of Proposals in NSF's
Small Business Innovation Research Program

The NSF uses a broad-based peer review process in all of its granting programs. The majority of these are very early-stage basic research programs that are not analogous to OST's typical project. However, one NSF program that has similarities to OST and is a good model for reviewing developmental programs is SBIR.[1] This program was established by Congress especially to support innovation by small businesses and to get these innovations to the end user. The SBIR program consists of three phases: Phase I determines scientific, technological, and commercial merit; Phase II develops the concepts further up to demonstration; and Phase III involves commercialization of the demonstrated technology (without funding support from NSF). Peer reviews are conducted on projects at both Phase I and Phase II.

The SBIR program employs a two-stage evaluation process. The first stage uses peer review to assess the technical merit of the proposals. Following an initial screening of all proposals by NSF staff (who ensure that they meet the minimum requirements for the SBIR program and decide which program unit should review each proposal), all proposals are peer reviewed by a group of qualified disciplinary scientists and engineers to assess their technical merit.

The second stage in the evaluation process is an examination by a group of experts from a commercial or applications perspective. These experts are selected on the basis of their expertise as representing the user or consumer perspective. Because SBIR programs are intended to deliver commercial innovations to the marketplace, this second review is designed to evaluate not the technical basis of the proposal, but rather the need for and acceptance of the innovation in the competitive marketplace. NSF program managers then use the results from both the peer review and the commercial review to decide which projects will be supported.

The most important element in the NSF SBIR review process is the combination of two distinct, early-stage evaluations to assess technical merit and end user applicability. NSF program managers have discovered that although the process takes two steps, it has enhanced the proportion of proposals that migrate from Phase I to Phase II as well as the possibility of commercial success, that is, acceptance by the end user.

[1]The SBIR program operates under Public Law 97-219 as amended by Public Law 102-5640 and is not unique to NSF. DOE, NIST, the Department of Commerce, and other agencies also have SBIR programs.

> **BOX 3.2**
> **Peer Review of Proposals in NASA's**
> **Office of Life and Microgravity Sciences and Applications**
>
> NASA's Office of Life and Microgravity Sciences and Applications (OLMSA) funds basic research and technology development related to life and microgravity sciences, in support of its mission to advance knowledge, improve the quality of life on Earth, and strengthen the foundations for continuing the exploration and utilization of space. OLMSA uses peer review in its grants award process to evaluate the scientific merit of proposals. OLMSA staff then evaluates the proposals for relevance to the program's objectives. OLMSA uses a contractor, Information Dynamics, Incorporated (IDI), to support the solicitation and peer review process, with subcontracted support by the Universities Space Research Association. IDI maintains a strong working relationship with NASA, but the review process remains independent.
>
> Peer reviews of research proposals are conducted by panels of experts selected by IDI on the basis of expertise, peer review experience, and panel balance. NASA may recommend reviewers and can exclude proposed reviewers if a reviewer has a known conflict of interest with the program (although this occurs only rarely). Prior to the review panel meeting, primary and secondary reviewers are assigned to each proposal. These reviewers (as well as any others who so desire) write evaluations of their assigned proposals and submit them to IDI. The peer review meetings at which proposals are discussed are closed so that reviewers are not influenced by external forces. At the review panel meeting, the panel as a whole assesses and ranks all proposals, and writes its final evaluation. The panel's evaluations and ranking are used as important input to NASA program managers in building their research program.

Society of Mechanical Engineers and the Institute for Regulatory Science (see Chapter 4).

PROJECT MATURITY EVALUATIONS

Evaluations of project maturity refer to peer reviews conducted as a project develops from a research idea to a technology that can be demonstrated and ultimately deployed. OST's current peer review process fits this review type. Another example of this type of review is summarized in Box 3.4. The objectives of the review, criteria for review, and panel expertise required should all change as a project moves through the maturation process, as the following examples note.

> **BOX 3.3**
> **Peer Review of Proposals by the American Institute of Biological Sciences**
>
> AIBS administers a variety of peer reviews in the areas of life and biomedical sciences for federal and state government agencies, as well as private institutions. To accomplish this, it uses various types of peer review, including mail reviews, panel meetings, and evaluations involving site visits. AIBS peer reviews have ranged from mail reviews of a single proposal to multipanel review of more than 2,000 proposals for the 1993 Army Breast Cancer Program. To identify potential peer reviewers, AIBS maintains a database of about 7,000 people recommended by members of AIBS and 42 affiliate societies. Conflict of interest and confidentiality are important considerations for the peer review program, and reviewers sign confidentiality agreements and conflict-of-interest statements. AIBS uses NIH and NSF standard rules for conflict of interest to avoid both real and perceived conflicts. AIBS peer review panel meetings are held in closed sessions to encourage frank and open discussion of proposals.

One important general point for all types of project maturity peer reviews is the need to distinguish clearly between the technical focus of *peer reviews* and nontechnical factors that play important roles in *decisions* to continue to support specific projects. Clearly, many nontechnical issues, such as public acceptability and relative DOE needs, should be considered by program managers when making decisions on which projects to support. These nontechnical issues should not be confused with appropriate criteria for peer reviews, however, because they could sidetrack reviewers into issues that are beyond their expertise or are difficult to resolve within the time constraints of a two- to three-day review. Program managers can incorporate such input into their decision-making process, however, in a variety of ways, including reviews (such as commercial viability reviews or "relevance reviews") conducted separately or in parallel with peer reviews (see Box 3.1). In the case of OST, the stage-gate reviews of its Technology Investment Decision Model (TIDM) provide such an opportunity to incorporate nontechnical input into the decision-making process.

In the following sections, the committee describes four specific stages of technology development in which peer review could be applied to evaluate the technical merit of a project. For each stage, the committee discusses the kinds of specific technical issues that could be of particular concern at this point of technology development. Also included is a description of some of the nontechnical factors that would be involved in decisions regarding which projects to support at each stage of development.

Entrance into the Applied Research Stage

Basic research may produce a concept thought to have relevance to solving a site-specific need. Program managers then may desire to see whether the concept has enough promise to justify moving into a program of applied research (e.g., Gate 1 of OST's TIDM; see Chapter 4 and Appendix A). A peer review as defined in this report could be very useful in assessing the technical merit of the concept. For example, is the science sound? Is it likely to develop into a technology to meet the stated need? Are other technologies already available? Addressing these types of questions would require peer reviewers who are experts in the relevant scientific and technological disciplines— individuals independent of those promoting the concept.

The decision to continue to fund such a project at this stage would consider other nontechnical factors in addition to the technical criteria used in peer review. For example, if favorable answers are given to the technical questions, is the technology likely to be developed in the time frame required by implementation schedules, with the funding and resources likely to be available to the program?

Entrance into Engineering Development

This is an important step in the technology development process because it often leads to the commitment of large amounts of funding for engineering development of a project. Such a review would address the technical adequacy of the technology—that is, whether all the scientific work needed to proceed to technology development is in hand and whether the technology is likely to work as promised to meet the need addressed.

The decision to continue to support a project at this stage (e.g., Gate 4 of OST's TIDM; see Chapter 4 and Appendix A) also would consider nontechnical aspects, such as regulatory performance standards and timetables, public acceptability, and DOE needs. Input on these nontechnical factors could be provided to program managers by other types of reviews, even reviews conducted in parallel to the peer reviews.

> **BOX 3.4**
> **Peer Reviews in DOD Strategic Environmental**
> **Research and Development Program**
>
> The Department of Defense's SERDP is the largest environmental research program within DOD (currently funded at $50 million to $60 million per year). SERDP is a statutory DOD–DOE–EPA science and technology program established to address the environmental needs of the DOD and congruent DOE needs. As such, SERDP responds to the specific, high-priority needs that are defined by the Deputy Under Secretary of Defense for Environmental Security. As a mission-oriented program, SERDP must ensure that the research it sponsors is both responsive to stated needs and of the highest scientific quality. SERDP is a DOD program, but it also addresses to some extent the strategic environmental needs of DOE and EPA, especially when there is some overlap with military needs. Research can be conducted by the uniformed services, DOE, EPA, and, starting in 1997, nongovernmental organizations (NGOs) on a competitive basis. SERDP employs peer reviews to evaluate proposals, ongoing projects, and its overall program.
>
> SERDP uses a multistage process to select proposals solicited from both the federal agencies and the private sector, and peer reviews are a critical part of this process. The process begins with Statements of Needs (SONs) solicited from the uniformed services. These statements are prioritized and generalized by the internal SERDP Technology Thrust Area Working Groups (TTAWGs; see below) for the four thrust areas of SERDP (cleanup, compliance, conservation, and pollution prevention), and a call for proposals is issued, either to government laboratories or to the broad research community (NGOs). The initial screening of full proposals from NGOs is performed by SERDP staff (and described below), whereas the initial screening of full proposals on the federal side is performed by SERDP's federal partners (DOD, DOE, and EPA), with each partner limited in the number of full proposals that can be submitted. Peer review is used in the evaluation of all proposals (from both federal and NGO communities).
>
> Proposals from NGOs are first evaluated by SERDP staff to ensure that the project addresses the SON and is appropriate for SERDP funding. Proposals are subjected to a mail peer review modeled after the system used by NSF. The peer review process is conducted by a support contractor under the control and supervision of SERDP staff. The contractor nominates potential peer reviewers, who must be approved by SERDP staff. Peer reviewers judge the proposals against a set of technical merit and personnel criteria and assign a numerical value from 1 to 4. Reviewers also are asked if they would fund each proposal based upon their overall assessment of the project and its cost. This review results in a relative ranking of the proposals within each SON.
>
> The proposals, along with peer review results, are then provided to the TTAWGs, which are responsible for recommending a program plan of projects to fund each year for each thrust area. TTAWGs are composed of scientific and technical personnel from DOD, DOE, and EPA, and are charged with evaluating the

NGO and federal proposals against a set of weighted evaluation criteria, which include technical merit, transition potential, quality of personnel, cost, and cooperative development (i.e., whether the cost is shared with other funding sponsors). In their review of proposals, TTAWG members consider peer review comments but are not constrained to follow the rankings provided by the peer review. TTAWGs are required to justify any recommendation that is not consistent with the peer reviewers' evaluation, however.

The recommended program plan, together with all of the results from peer reviews, is presented to the SERDP Technical Director and Executive Director for review and final selection. A final review of the SERDP program is performed by the SERDP Scientific Advisory Board (SAB).[1] The SAB must approve all new projects and review all project renewals above $900,000 per year; it also conducts an assessment of the overall SERDP program. During their review, SAB members have access to all of the data from both the peer review and the TTAWG review. SAB rejected projects at the rate of 19 percent in fiscal year 1997 (DOD, 1998).

Even though there is a broad knowledge base in the TTAWGs and the SERDP staff, the use of peer review to evaluate proposals and ongoing projects, as well as to assess the overall balance of the SERDP program, ensures that each SERDP project is evaluated by independent experts in the field. This is vital to maintaining the scientific and technical quality of SERDP. Peer review is utilized as a tool that aids in the proposal selection process as opposed to being a decision process unto itself.

---

[1] The act governing SERDP (10 USC 2901-2904; Public Law 101-510) as amended in 1997 mandates a Scientific Advisory Board (SAB) to review all projects funded above $1 million per year and to provide other appropriate advice to the SERDP Council. The SAB is chartered under the Federal Advisory Committee Act (FACA) and its members represent a balanced mix of disciplines, including engineering, chemistry, oceanography, ecology, and health sciences. Although such a standing FACA panel may not be feasible for OST in the short term, many of the above concepts could be applied on an ad hoc basis.

Entrance into the Demonstration Stage

This step represents a decision on whether or not to proceed to full-scale, on-site demonstration of a technology (e.g., Gate 5 of OST's TIDM; see Chapter 4 and Appendix A). At this point the science should have been fully verified, and technical questions would have more to do with whether the engineering development test data are adequate to design a full-scale demonstration that will be safe and cost-effective.

Decisions to continue to support a project at this stage also would consider nontechnical factors similar to those discussed above for other project maturity evaluations. In particular, questions of public acceptability might be given high priority because a decision to proceed with a technology at this stage will result in a full-scale project operating at a contaminated site where public sensitivity is likely to be high. Thus, a program manager might solicit input from experts knowledgeable on risk as well as particular site stakeholders' perceptions of risk to supplement the results of a peer review. Such stakeholders would not be appropriate peer reviewers, however.

Predeployment Reviews

At this stage of development the technical soundness of the concept should have been proven, but the efficacy, cost-effectiveness, safety, and regulatory acceptability of the technology may have yet to be demonstrated. A predeployment peer review would examine the adequacy of the test data generated during the technology's development—for example, whether these test data show that the technology is safe and cost-effective.

Decisions to continue to support a project at this stage (e.g., Gate 6 of OST's TIDM; see Chapter 4 and Appendix A) also would involve nontechnical considerations, including regulatory compliance, intellectual property, liability, public acceptability, risk, and safety. A program manager could receive input from experts in these areas, and such input could take the form of separate reviews conducted in parallel or separate from peer reviews on technical matters.

PROGRAM BALANCE EVALUATIONS

If a fully assessed set of needs is available, management might want to obtain an independent assessment of program balance, that is, whether the technology development program adequately addresses those needs, given the resources available. For such an evaluation, the review panel as a whole should have technical expertise covering the entire scope of relevant technologies. Reviewers also should have knowledge of other existing technologies or of technologies in development nationally and internationally. To avoid conflicts of interest the panel should be independent of the managers who constructed the existing program and PIs connected with the program.

In OST's case, this would be achieved by going outside OST and those it funds. If one purpose of such an evaluations were to uncover duplication or potential synergies with projects in other parts of DOE, it might be advisable to have some panelists who are familiar with all relevant DOE programs, perhaps DOE employees, or to have such persons present at the review session to advise panel members. The committee cautions that extreme care should be used in appointing DOE employees as panelists because of the potential for conflicts of interest (see Chapter 5). Three examples of the application of peer review to evaluate program balance are given in Boxes 3.5, 3.6, and 3.7.

NEEDS DETERMINATIONS

A peer review of program needs could assess, in OST's case for example, research and development needs to address environmental problems at contaminated sites. These problems are the fundamental drivers of the technology development program—that is, the technologies are developed to solve the problems caused by contamination. A needs review of the whole program might be conducted to examine the entire suite of needs that the program must address in order to begin to assess program balance. It can be argued that a needs review is not necessary because the personnel at the sites are closest to the problem and therefore require no review. However, this very closeness when coupled with funding pressures defines a situation in which peer review can be beneficial. An example of a needs evaluation used in industry is provided in Box 3.8.

For such a review the technical expertise required would be that necessary to assess both the types of contamination and the peculiarities of specific sites, and thus would be a very broad scope of expertise. In addition, panels would have to be constructed to ensure a fair and balanced assessment of all needs—that is, to avoid having members with interest in giving a particular site inappropriate priority for technology development. For example, a member of the public from a community affected by a DOE site should not be a panel member due to potential conflicts of interest or bias.

> **BOX 3.5**
> **Peer Review of Programs at National Institute of Standards and Technology Laboratories**
>
> Since 1959, NIST (and its predecessor agency, the National Bureau of Standards) has practiced external assessment of its activities, specifically by technical panels of the NRC's Board on Assessment of NIST Programs. The longevity of this practice and the NRC's ability to attract volunteers for service on its panels attest to its success and its value to NIST. Elements of the NIST process may apply to the OST, but some salient differences between the situations of NIST and DOE–OST should be noted.
>
> NRC panels review the NIST Measurement and Standards Laboratories, each of which has responsibility for a discipline area (e.g., chemistry, physics, materials, information technology) related to NIST's overall mission in measurements and standards. NIST programs typically continue for decades although individual projects come and go within programs. Although NIST focuses on industry and commerce, research also relates to defense, health, environment, space, and science. The problems NIST addresses are usually technically narrower and better defined than those faced by DOE–OST. For example, NIST might well develop a standard measurement method to determine very low levels of impurities in semiconductor wafers, but not methods to lower the level of impurities. NIST's typical client is technically sophisticated, although this is not always the case. Finally, having no regulatory authority, NIST typically has a cooperative relationship with its clients. There are at least two relevant similarities between NIST and OST: (1) the problems appropriate to its mission and needing attention are too great for the resources available, so priorities must be set; and (2) these problems span many fields of science and technology.
>
> NIST meets the criteria defining peer review by asking the NRC to run its assessment system. Because the panels are constituted by and report through the NRC, they are independent and external. The NRC appoints panel members and controls the production of the official reports to NIST. The NRC is able to access the wide range of technical experts needed to review NIST programs. The National Academies of Sciences and Engineering provide a good range of contacts and certify expertise. A great many panel members are from industry, thus involving stakeholders and individuals with practical experience with the problems being addressed. The panels are "balanced" so that no industry or company dominates. Panel members serve overlapping three-year terms, with one renewal possible, to allow for continuity and follow-up on recommendations. This also allows for increasing understanding of the programs being reviewed. Dialogue between NIST officials and the NRC occurs throughout the review process to facilitate the process and maximize its usefulness to NIST.

BOX 3.6
Peer Review of Programs of the U.S. Army Engineer Waterways Experiment Station

One way of supporting the Government Performance and Results Act (GPRA; Public Law 103-62) of 1993 is to conduct peer reviews of programs at a departmental or laboratory level (e.g., about 200 professional staff) and address the quality of staff, research, products, and facilities. This approach is used at the U.S. Army Engineer Waterways Experiment Station (WES; Conway et al., 1996, 1997).[1] Over a two- to four-day period a panel of three to five technical leaders reviews programs and biographical documents, receives technical presentations, views facilities, and interacts with staff. A 10- to 15-page report is prepared that describes the review procedure, generalizes findings, and—for each of the four review elements—lists strengths, areas for improvement (usually with suggestions), and a numerical rating from 1 to 5 (poor to excellent). The panel, with partial membership rotation, repeats the review on a one- to two-year cycle to assess progress in areas identified for improvement and to identify new areas to be addressed.

Each panelist in the WES review has demonstrated technical expertise and leadership at the international level. The panelists' expertise is complementary: for example, in the review of a large environmental laboratory, the panel consisted of an environmental engineer, a bioengineer or scientist, and an environmental biologist. Panelists make recommendations that have the potential for broad impact, are not too numerous, and largely can be addressed in the short term. Particular attention is paid to Congress's concern over the need for performance metrics, for detailed strategic plans and milestones, for coordination across agencies, and for impact in the short term (American Institute of Physics, 1997).

This model could be used by OST to review a group of similar projects under a focus area.

---

[1] NIST also currently uses this approach.

BOX 3.7
Peer Reviews of Research Proposals and Programs in DOE's Office of Basic Energy Sciences

DOE's Office of Basic Energy Sciences routinely carries out peer reviews of both proposals and programs. BES peers reviews include the following:

1. *BES programs at national laboratories (e.g., Lawrence Berkeley Laboratory, Argonne National Laboratory).* Combined reviews of specific programs (overall research performance, relevance, future directions) and individual research projects are performed at least annually at each laboratory site. Specific review procedures differ from lab to lab, depending on the nature of the research conducted. For example, Lawrence Livermore National Laboratory has special procedures for handling its classified programs. The laboratory director, in consultation with the BES division heads and program managers, chooses the peer review committee and chair from the scientific community after a list of candidates is solicited from PIs. DOE headquarters may also suggest potential peer reviewers. The criteria for selection of committee members include expertise and eminence in the fields of research, and committees generally include nationally and internationally respected scientists. DOE headquarters staff may also attend the review.

In preparation for the site review, program managers, directors, and PIs provide written summaries of their research including the annual field work proposals, publications from their research, updated curricula vitae, and other relevant data to reviewers and DOE managers. Following the reviews, the peer review committee chair meets privately with the laboratory and division directors to discuss the review, as well as future directions. After discussions with the peer review committee, the division heads send recommendations (including positive and negative comments) to the individual investigators. The procedures are summarized as follows:

- The Peer Review Committee presents oral and written reports to the laboratory director and division heads.
- The laboratory director provides oral and written summaries for division heads and PIs.
- The division heads and PIs respond in writing to peer review and demonstrate adjustments to the programs, if needed. Failure to adjust to BES program shifts and directives could lead to loss of funding.

2. *Individual Proposals (not affiliated with national laboratories).* BES handles these in essentially the same manner as NSF does in peer reviews of grant proposals (see NSF, 1995, 1997). Proposers follow the criteria from DOE's request for proposals and submit proposals to the appropriate office, which sends them out for peer review and written evaluations. DOE uses the reviews to aid in project selection project selection and funding decisions.

3. *BES national reviews.* DOE–BES occasionally holds peer reviews of individual and laboratory research programs on a national/international basis. Such reviews generally are held in Washington, D.C., and reviews by BES managers occur simultaneously. The principal objectives of these are to ensure that the research projects are progressing optimally and to maintain the highest scientific and technical standards.

BOX 3.8
Needs Evaluation at Chiron Corporation

Chiron Corporation, headquartered in Emeryville, California, near San Francisco, is a biotechnology company that combines diagnostic, vaccine, and therapeutic strategies for controlling disease. Chiron participates in three global markets: (1) diagnostics, including immunodiagnostics, critical care diagnostics, and new quantitative probe tests; (2) pediatric and adult vaccines; and (3) therapeutics, with an emphasis on oncology and infectious diseases. Chiron has research programs under way in gene therapy, gene transfer, combinatorial chemistry, and recombinant proteins targeted toward oncology and cardiovascular and infectious diseases.

As a research-driven company, Chiron is especially sensitive to the thorough review of various parts of its research portfolio. Although the majority of the regular program reviews are performed internal to the corporation, a peer review process using external experts not associated with the company has been invoked at critical stages in several programs. One of the most extensive reviews was done prior to launching a major new program in the Chiron Diagnostics business unit.

The new program was conceptualized and developed internally over a period of almost a year. No investments in new products were made, but considerable effort was expended in market review, scientific assessment, and product planning. The undertaking was of considerable scope and called for a significant multiyear investment. As plans were formalized, additional internal resources were engaged; however, no major capital investments were made. The final stage in the initiation of this business thrust was a peer review session in which approximately ten outside experts were engaged for two days to review and critique the plans. The experts were also asked to make more extended comments on a number of relevant areas covered by the plans. The proceedings were recorded for further reference and review by Chiron staff not in attendance.

The result of this intense external review was a significant change in the direction and magnitude of the program. The expert reviewers were able to provide an alternative view of the cost and time required to mount a competitive program given the Chiron starting point. Likewise, they gave valuable input on the competitive landscape of the field and the likelihood of rapid acceptance of the products in clinical medicine. In this case, Chiron responded carefully to the external reviewers' recommendations and was able to adjust the deployment of financial and human resources into areas of greater short-term impact. Like OST, Chiron needed to address a specific problem under constraints (in this case, governed by the market) such as budgets and schedules. However, Chiron recognized the value of peer review, and, before investing large amounts of funds, used peer review to improve its investment.

4

OST's Peer Review Program

In this chapter, the committee presents a description of OST's peer review program as of April 1998. Much of the information was based on *Implementation Guidance for the OST Technical Peer Review* (DOE, 1998b), ASME's *Manual for Peer Review* (ASME, 1998), RSI's Handbook of Peer Review (RSI, 1998), as well as presentations by OST, ASME, and RSI staff. The committee's analysis of the current state of the peer review program is presented in Chapter 5.

OST's peer review program was designed to provide program managers with credible, independent evaluations of the scientific and engineering merits of its technology projects. The peer review program evaluates technology projects at various stages of development (from basic research to late-stage demonstrations and implementation of the technologies)—not solely at the proposal stage, as practiced by many other government agencies. OST established the peer review program to provide one important input to its focus area/crosscutting area (FA/CC) program managers as they make "go or no-go" decisions on technology projects supported by OST.

TECHNOLOGY INVESTMENT DECISION MODEL

OST's Technology Investment Decision Model is a procedure OST developed to provide a common basis on which to assess and manage the performance, expectations, and transition of technologies through the development process (Paladino and Longsworth, 1995). It is a user-oriented decision-making process for managing technology development and for linking technology development activities with cleanup operations (see Appendix A for a more complete description of the TIDM). It should be noted that the TIDM procedure has been documented by OST, and OST does use the overall TIDM framework to track its projects, but OST does not yet use the TIDM approach in its decision making consistently across the organization.

The TIDM identifies seven R&D stages from basic research through implementation of a technology (see Figure 4.1). At each stage, specific criteria, requirements, and deliverables form a common basis for technology assessment. In the model, stages are separated by "gates"—decision points at which projects are evaluated for funding of the next stage. The stage-gate process is meant to provide for evaluation of projects at all stages of development against technical and nontechnical criteria selected to ensure that the technologies developed will provide superior performance, will meet the acceptance requirements of the intended customers, and can be moved into the marketplace.

At each gate, OST's FA/CC program managers are responsible for evaluating a technology's documentation in accordance with the appropriate criteria. If the FA/CC program manager determines that the technology warrants passing through a gate, the technology maturation process continues. If the program manager determines that the technology does not warrant further consideration, then funding is discontinued. See Appendix A for a description of individual stages and gates of the TIDM.

Peer reviews are intended to be used by OST in the TIDM to provide an independent, external evaluation of the technical merits of a technology. These evaluations are intended to assist decision makers in making decisions regarding further support for technology projects as they proceed through OST's TIDM. In part, because the TIDM evaluates projects against both technical and nontechnical criteria, OST's peer review program does not directly make decisions regarding the funding of OST technologies. Rather, peer review results are intended to be one input into the decisions made at the critical decision points (gates) of the research and development process.

ROLES AND RESPONSIBILITIES

As noted above, the main decision makers in terms of funding technology projects within OST are the individual FA/CC program managers. As such, FA/CC program managers are the main "customers" of the OST peer review program. The peer review program is managed within OST by the Peer Review Coordinator, who represents the Deputy Assistant Secretary for Science and Technology. The peer review sessions themselves are conducted by the American Society of Mechanical Engineers, with administrative and technical

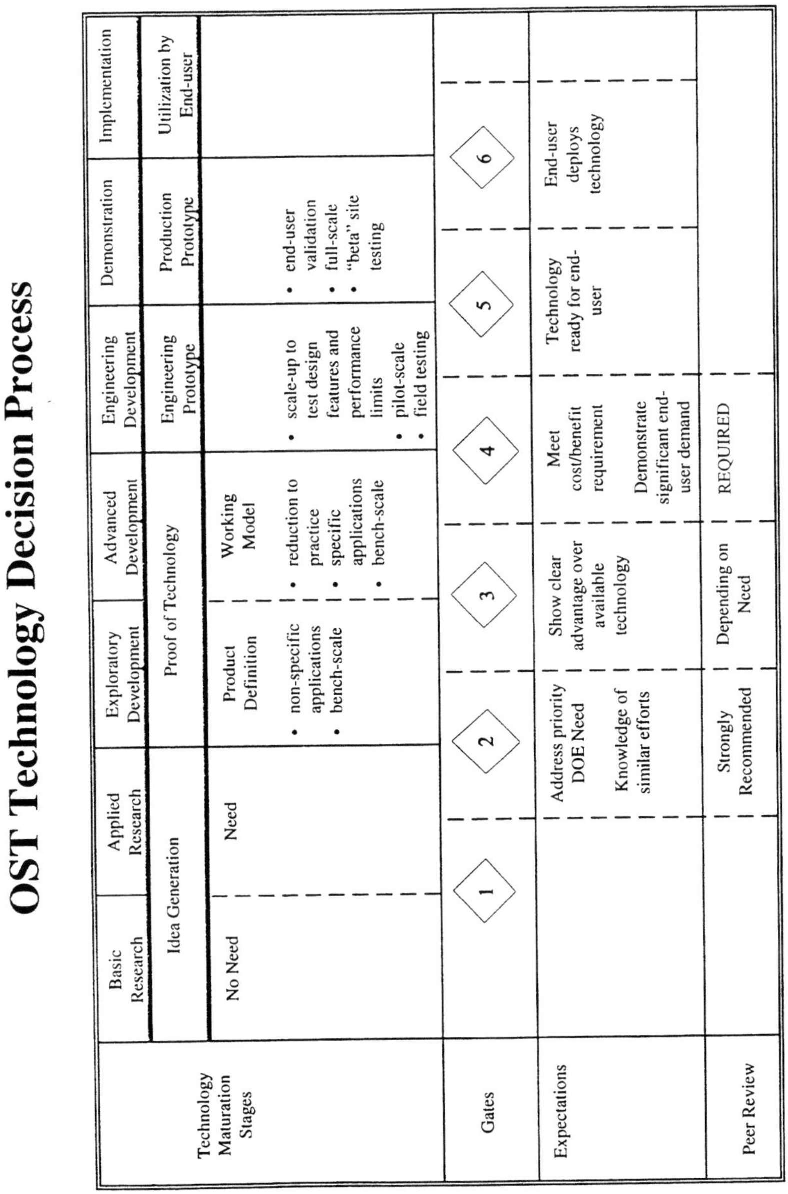

OST Technology Decision Process

Technology Maturation Stages	Basic Research	Applied Research	Exploratory Development	Advanced Development	Engineering Development	Demonstration	Implementation
	Idea Generation	Need	Proof of Technology			Production Prototype	Utilization by End-user
	No Need		Product Definition	Working Model	Engineering Prototype		
			• non-specific applications • bench-scale	• reduction to practice • specific applications • bench-scale	• scale-up to test design features and performance limits • pilot-scale • field testing	• end-user validation • full-scale • "beta" site testing	
Gates	1	2	3	4	5	6	
Expectations		Address priority DOE Need Knowledge of similar efforts	Show clear advantage over available technology	Meet cost/benefit requirement Demonstrate significant end-user demand	Technology ready for end-user	End-user deploys technology	
Peer Review		Strongly Recommended	Depending on Need	REQUIRED			

FIGURE 4.1 Diagram of OST's Technology Investment Decision Model (DOE, 1998b).

support provided by the Institute for Regulatory Science. Funding for the review program is provided directly to RSI through a $1.2 million annual grant from OST. According to OST, ASME was chosen to conduct parts of OST's peer review process because of its long history as a technical professional society, its work in developing and categorizing technical and engineering codes and standards, its experience conducting peer review for other organizations, and its independence from DOE. This section describes the primary roles and responsibilities of the various entities involved in the OST peer review program.

Peer Review Coordinator

The Peer Review Coordinator (under DOE's Chicago Operations Office) is the principal federal official responsible for managing the peer review program's day-to-day activities. Specific responsibilities include developing a targeted plan for the program, including a prioritized list of projects to be reviewed; receiving, processing, and scheduling peer review requests from FA/CC program managers; coordinating peer review activities among FA/CC program managers, the ASME Peer Review Committee (see below), and the review panels; ensuring that reviews are executed in a timely manner; ensuring that FA/CC program responses to review recommendations are included in Final Reports; managing the budget and records for OST; and tracking activity metrics for the program (DOE, 1998b).

Headquarters Peer Review Program Manager

The headquarters Peer Review Program Manager is responsible for monitoring all peer review activities within OST. The program manager also works with the Peer Review Coordinator to provide direction on policy, program planning, and budgetary issues.

FA/CC Program Managers

The FA/CC program managers work with the Peer Review Coordinator to develop a prioritized list of technologies to be reviewed and initiate the peer review process by making written requests for peer reviews to the OST Peer Review Coordinator. FA/CC program managers also provide documentation to reviewers, prepare responses to the Report of the Review Panel, coordinate the responses with the director of the Office of Technology Systems, and cover the cost of FA/CC program personnel and material needed for the peer reviews.[1] The FA/CC program managers are responsible for incorporating the results of peer reviews into the decision-making process.

Principal Investigators

The principal investigators of a project are the scientists and engineers who perform the technology development work. Each project's PIs are responsible for providing relevant technical background information for the peer reviewers 30 days prior to the review (See Table 4.1). For reviews of Types I and II (see below), the PIs are also responsible for presenting and responding to the review panel.

American Society of Mechanical Engineers

The ASME establishes and sanctions the Peer Review Committee and its Executive Panel through formal approval by the Center for Research and Technology Development of ASME's Council of Engineering. The ASME ensures that peer reviews follow ASME procedures and serves as a resource for identifying potential peer review panel members. As well as being able to provide names from its own membership, the ASME will attempt to reach collegiality agreements with other relevant professional societies, as necessary, to identify additional potential nominees.

ASME Peer Review Committee and its Executive Panel

The Peer Review Committee (PRC) is a standing body of the ASME whose sole purpose is to oversee OST's peer review program and enforce relevant ASME policies, including compliance with professional and ethical requirements. The PRC is a consensus body that meets several times per year and consists of members chosen for their competences and diversity of views. Thirteen members currently serve on the PRC (see Box 4.1). Membership in ASME is not required for appointment to the PRC (except for members of its Executive Panel—see below). The PRC is responsible for appointing persons

---

[1] At present, the direct cost of the peer review itself is covered by the grant to RSI, rather than by the budget of the FA/CC program manager requesting the review; however, the costs of preparing for the reviews and presenting to the review panel are covered by the FA/CC program.

nominated by the Executive Panel to individual peer review panels and approving Interim Reports (see section "Review Reports" below) before they are issued as ASME-sanctioned public documents.[2] The PRC also reviews and approves the Annual Report and presents it, along with a comprehensive evaluation of the technical quality of the OST program, to the Deputy Assistant Secretary for OST at an annual meeting.

The Executive Panel (EP) of the PRC consists of three to five ASME members who typically have served in leadership positions within ASME (i.e., president, vice-president, division chair). The EP is responsible for overseeing the day-to-day operation of the peer review program and acts on behalf of the PRC between its meetings. The EP meets approximately four times per year but conducts most of its business by correspondence.

The first annual meeting of the ASME Peer Review Committee was held in Washington, D.C., on November 24, 1997, at which time the first Annual Report (ASME, 1997) was released. The report includes a description of the program, recommendations addressed to OST regarding the conduct of reviews, and a compilation of all the peer review reports produced in Fiscal Year 1997.

Peer Review Panels

Peer review panels consist of three or more technical experts chosen for their knowledge of the specific technology to be reviewed. The complexity of the technology being reviewed and the type of review are used by the PRC to choose the number of experts needed on a particular panel. The peer review panel conducts the review, prepares a Report of the Review Panel (formerly termed Consensus Report) detailing its recommendations and observations, and transmits the written report to the Peer Review Coordinator, who forwards it to the FA/CC program managers for response. In most cases, peer review panels produce a single report that summarizes the consensus views of all members. In cases where a consensus among all members is not reached, however, the Report of the Review Panel can include a minority opinion as an appendix (see discussion of Review Reports below). Peer review panels are terminated upon completion of their specific review task.

---

[2]As part of its assessment, the PRC may add comments to the Interim Reports before issuing them in final form.

> **BOX 4.1**
> **ASME Peer Review Committee**
>
> Charles O. Velzy (chair), consultant[1,2]
> Gary A. Benda, NUKEM Nuclear Technologies, Corp.[2]
> Erich W. Bretthauer, Bryce Meadows Development Corporation
> Ernest L. Daman, Foster Wheeler Development Corporation[1,2]
> Robert A. Fjeld, Clemson University
> John T. Greeves, U.S. Nuclear Regulatory Commission
> William T. Gregory III, Foster Wheeler Environmental Corporation[2]
> Nathan H. Hurt, IDM Environmental Corp.[1,2]
> Peter B. Lederman, New Jersey Institute of Technology
> Jeffrey A. Marqusee, Office of the Deputy Under Secretary of Defense for Environmental Security
> A. Alan Moghissi, Institute for Regulatory Science[1,2]
> Goetz K. Oertel, Association of Universities for Research in Astronomy, Inc.
>
> **Staff**
> Howard E. Clark, American Society of Mechanical Engineers
>
> ---
> [1] Member of Executive Panel
> [2] Member of ASME

Institute for Regulatory Science

OST chose RSI to provide administrative and technical support for peer review activities. RSI's responsibilities include meeting planning, compiling and distributing background materials for members of review panels, facilitating peer reviews, assisting in the identification of potential review panel nominees, and providing a Technical Secretary who participates in all meetings of the review panels and assists in the preparation of their technical reports. The RSI President manages the review program, sits on the ASME EP, and supervises RSI's activities related to the review program.[3]

---

[3] Because the committee was asked not to review the DOE–ASME–RSI relationship, it could not arrive at a complete understanding of this relationship and therefore makes no comment on it.

SELECTION OF TECHNOLOGIES TO BE REVIEWED

OST policy requires that a technology peer review be conducted on each technology or system. In response to this committee's interim report (NRC, 1997b), OST recently adopted a new policy requiring that candidate technologies for peer review be identified as those that fulfill at least one of the following three project selection criteria (not in priority order):

1. Gate 4: technologies or systems passing into the engineering development stage; or
2. More than three years of funding: projects that have been funded for more than three years and have not been peer reviewed; or
3. New start: projects that were initiated in FY97 or FY98 and proposed new starts.

In planning for peer reviews to be conducted in FY98, OST used its Technology Management System Database (consisting of more than 800 technologies or systems that have been funded by OST from FY89 to FY98) to identify more than 200 technologies that meet at least one of these three criteria. This list was sorted by the amount of funding each project had received to date. The list was then provided to all FA/CC program managers so that they could further prioritize their list of projects by the dates at which peer review results would be most valuable for making programmatic and/or funding decisions (DOE, 1998b, p. 11). Ultimately, a plan to review 38 specific technologies during FY98 was prepared.

DOCUMENTATION REQUIRED FOR REVIEW

OST's Implementation Guidance (DOE, 1998b) lists the written documentation generally required for a peer review at each stage of development of a project (Table 4.1).

TABLE 4.1 Documentation Required by OST for Peer Reviews Conducted at Different Gates of its Technology Investment Decision Model.

Documents	Gates					
	1	2	3	4	5	6
Proof of principle	✓	✓	✓	✓	✓	✓
Literature references	✓	✓	✓	✓	✓	✓
Progress report (topical)	✓	✓	✓	✓	✓	✓
Needs document		✓	✓	✓	✓	✓
Test plan at the appropriate scale		✓	✓	✓	✓	✓
Data quality assurance plan			✓	✓	✓	✓
Proof of design				✓	✓	✓
Construction plan					✓	✓
Implementation plan						✓

SOURCE: DOE, 1998b, p. 18.

TYPES OF REVIEWS

The size and scope of each review panel depends on the specific technology(ies) being reviewed and the areas of expertise required to address the review criteria. In general, there are three types of peer reviews:

- Type I: multitechnology reviews of a complex nature (five or more reviewers);
- Type II: in-depth reviews of a single technology (three or more reviewers); and
- Type III: reviews of documents requiring no meeting of the review panel (i.e., mail review; three or more reviewers).

To reduce costs and increase the number of projects that can be reviewed, OST now reserves Type II reviews for emergency situations where a program manager requires a short-term evaluation of the technical merits of a specific project or proposal, and encourages peer reviews of similar technologies to be grouped together in Type I reviews whenever possible (DOE, 1998b).

SELECTION OF REVIEWERS

Selection Criteria

Peer reviewers are selected by three generally recognized criteria: (1) education and relevant experience, (2) peer recognition, and (3) contributions to the profession. A minimum of a B.S. in a relevant engineering or scientific field and significant experience in the technical area being reviewed is required for participation on a review panel. Peer recognition is assessed by measures such as activity in professional societies and scholarly organizations. Contributions to the profession include publications in peer reviewed journals, patents, and meeting presentations.

Panel Formation

According to presentations by RSI staff at the ASME Peer Review Committee meeting in November 1997, ASME uses three electronic databases and the ASME membership list (approximately 120,000 names) to identify reviewers meeting the above criteria. In addition to a database of those who have previously served as reviewers (approximately 130 names), ASME and RSI have a working database of nominated reviewers that contains approximately 1,000 entries and a list of potential reviewers with approximately 5,000 entries. RSI uses these databases to identify potential reviewers who then must be approved by the ASME PRC. Panel members are reimbursed for travel expenses and generally receive an honorarium of several hundred dollars per review.

Conflict of Interest

Individuals with any real or perceived conflicts of interest with respect to the subject of the review may not serve as members of a review panel. According to ASME, "an individual who has a personal stake in the outcome of the review may not act as a reviewer" (ASME, 1997, p. 8). The ASME PRC has interpreted this provision as excluding all employees of DOE and DOE-owned facilities operated by contractors, including national laboratories. Potential conflicts of interest among members of the review panels, the PRC, and RSI staff are handled by requiring all who participate in the review program to sign a statement certifying that they do not have personal and financial interest in the outcome of the review. The ASME conflict-of-interest policy further requests that reviewers and members of the PRC and its Executive Panel recuse themselves from deliberations on any matter in which there may be an appearance of a conflict of interest.

REVIEW CRITERIA

OST established the following general review criteria (DOE, 1998b):

1. *Relevancy:* Is the technology applicable to the specific contaminants or conditions as claimed?
2. *Scientific and technical validity:* Is the technology consistent with established scientific, engineering, and industry principles and standards?

- Is the likelihood of its success reasonable?
- Are the data reliable?
- Are the stated scientific deductions meritorious?
- Are the project personnel technically qualified?
- Are the best available R&D practices used?

3. *Nonduplicative or superior to alternatives:* Is the technology duplicative or inferior to existing technology?
4. *Data validity:* Is the data collection process complete and valid, and can results be applied to regulatory, cost-benefit, stakeholder, and risk evaluations?

OST has provided no guidance on the relative weighting of these general criteria. However, OST policy states that these general review criteria must be augmented by technology-specific criteria. For each review, the FA/CC program manager submits a preliminary list of technology-specific criteria to the Peer Review Coordinator, who forwards them to RSI in the written request that initiates the review process. The proposed criteria are discussed among the Technical Secretary, Peer Review Coordinator, and OST managers, as necessary. Final criteria are approved by the EP of the ASME Peer Review Committee. These criteria are used by PIs and other presenters to organize written materials and oral presentations for the peer review. In addition, the review panel does have the authority to pursue other issues that arise during the review.

REVIEW SESSION

The peer review session for Type I and II reviews consists of five parts:

1. *Introduction.* A representative of the PRC describes the review program and discusses the review criteria, and the Peer Review Coordinator describes the OST process.
2. *Description of the project.* PIs give presentations describing the details of the technology under review. Stakeholders may present technical aspects of their concerns. Audience members and the review panel may ask questions of the presenters.
3. *Executive session.* The review panel meets in closed session to identify unresolved issues, unanswered questions, and additional needs for information.
4. *Discussion (questions and answers).* The review panel may ask questions of the presenters and other proponents of the project to resolve the questions raised in closed session.
5. *Executive session.* The review panel writes its report in closed session with the assistance of the Technical Secretary. This session commonly lasts half a day.

Throughout the open sessions, presenters are given the opportunity to respond to questions and clarify information provided to the panel. In many cases, presenters and other proponents of the project have been allowed to provide additional information or background materials to the review panel, on the condition that it is presented before the panel completes its deliberations. All of the peer review session is open to the public, with the exception of the executive sessions. The chair of the review panel presides over the entire session.

For Type III reviews, the format of the review is similar, but occurs by mail and phone. The introduction of the review is conducted via conference call. The review panel is generally given one to two weeks to read the written material being reviewed and to prepare comments. The Technical Secretary consolidates the comments and prepares the Report of the Review Panel, which is provided to the panel members for their review and comment. An executive session is held via conference call to discuss unresolved issues before the report is submitted to the Peer Review Coordinator.

REVIEW REPORTS

The outcome of each review is contained in a technical peer review report, which is prepared in three phases: Report of the Review Panel, Interim Report, and Final Report. The Report of the Review Panel is prepared at the review session with the participation of all panel members and the assistance of the Technical Secretary. According to OST policy, it should typically contain the following items (DOE, 1998b):

- introduction;
- project summary;
- objectives of the review;
- evaluation criteria;
- findings of the panel;
- recommendations of the panel;
- appendix including any minority reports or extra information that may seem unnecessary for the report but be of potential interest to readers;
- biographical summaries of the members of the review panel; and
- a list of references used in the analysis.

The Report of the Review Panel is provided to OST investigators and managers who are requested to notify the Technical Secretary of any errors within 5 days and to respond in writing to the recommendations within 30 days. The response is required to be clear and detailed and to include a schedule of implementation, if possible. A summary of the responses containing their salient features is added to the Report of the Review Panel to form the Interim Report. After DOE's response, the ASME PRC reviews the Interim Report, and may add to it substantive, explanatory, clarifying, and supplementary comments and recommendations. It is this committee's understanding that the PRC may not alter the sense of the review panel's report. The Interim Report as reviewed and approved by the PRC is issued as the Final Report.

Copies of the Reports of the Review Panels, the responses from the investigators and managers, and the Interim Reports are made available to members of the PRC for review and comment during the entire review process. At the end of the fiscal year (or after a specific period), an Annual Report is prepared by the PRC. This contains Final Reports of all projects, recommendations of the PRC, formal letters to OST,[4] and other information that,

---

[4] At its January 26, 1998, meeting the PRC voted to send several formal letters to OST, in order to provide more timely recommendations prior to the next Annual Report.

in the judgment of the PRC, would be beneficial to OST. This report is printed and widely distributed to FA/CC managers, PIs, and others.

FEEDBACK OF PEER REVIEW RESULTS INTO PROGRAM MANAGEMENT AND DEVELOPMENT DECISIONS

OST's TIDM requires that peer review reports be used in all funding decisions made at Gate 4 and any subsequent gate in the technology development process. In the gate reviews, these reports are intended to constitute one input to be considered in determining whether to continue funding for a project. As discussed in Chapter 5, OST has not yet demonstrated that the results from this new peer review program are used consistently in its decision-making process.

5

Analysis of OST's Peer Review Program

This committee published its interim report (NRC, 1997b) in October 1997. At that time, the committee found that OST had made progress in the implementation of its peer review program. The committee recognized that OST had developed a process for selecting reviewers, developing technology-specific review criteria, and executing peer reviews. Despite the relatively small number of reviews that had been conducted at that time, the committee concluded that OST's peer review program had the potential to be fair and credible. The committee also recognized a number of key obstacles that would have to be addressed before OST could fully achieve the objectives of this program, however. In particular, the committee noted problems with some of the procedures actually used to plan for and select projects to be reviewed, the general review criteria, and the application and usefulness of the results of the peer reviews. In this chapter, the committee examines the status of OST's peer review program by considering the main components of the peer review process.

The analysis is based on the committee's review of OST documents, presentations by DOE staff, observations of recent peer reviews and ASME Peer Review Committee meetings, and the committee's assessment of peer review reports. In each section of this chapter, the committee revisits the recommendations from the interim report and discuss specific changes that OST has implemented since the publication of its interim report to address these issues. Relevant aspects of OST's current peer review program and the OST organization are described in Chapter 4.

DEFINITION OF PEER REVIEW

In the past, OST has used the term "peer review" generally to refer to internal reviews of OST projects by qualified EM technical staff who were not involved directly in the specific project under review. This use of terminology caused confusion and misunderstanding within both OST and external review groups (e.g., GAO, NRC) who continued to criticize OST for lack of a credible peer review program.

The committee believes that at least part of the criticism leveled at the OST project review process has resulted from inconsistent and inaccurate descriptions of the processes involved (e.g., internal peer review, technical peer review). In the interim report, the committee therefore recommended that "OST should restrict the term 'peer review' to only those technical reviews conducted by independent, external experts. OST should adopt alternative terms, such as 'technical review,' for its internal reviews of scientific merit and pertinency" (NRC, 1997b, p. 10). In response to the committee's interim report, OST recently issued a policy restricting the term peer review to only those technical reviews conducted by independent, external experts (DOE, 1998b).

BENEFITS OF PEER REVIEW

The OST technology development program involves the expenditure of hundreds of millions of increasingly scarce federal discretionary dollars each year. It is particularly important that decisions about investment of funds are based on sound criteria (both technical and nontechnical) and that the decision-making process is respected by all parties in the technology development program. By ensuring that technologies are held to the highest technical standards, rigorous peer review can be an important tool for OST in meeting these objectives, as well as improving those projects that are funded. Thus, peer review at an early stage can ensure that projects are technically sound before they are developed (and funded) so that they will be able to achieve customer satisfaction. If they are not technically sound, they will not satisfy the customer.

The question also has been raised as to whether an assessment or expression of customer satisfaction should or could be a surrogate or substitute for peer review. Customer satisfaction and technical merit (as judged by independent peer review) are separate issues—each is important, but neither is sufficient to itself. In general, customers will not necessarily have knowledge of the technical aspects of the technology in question or of alternatives to it. Even when there is great expertise resident in the customers, it is nonetheless useful (e.g., in the federal funding process) to have an alternative judgment provided by disinterested parties. By the same token, a high-quality peer review is no substitute for customer satisfaction. Regardless of the peer review results, the customer should have confidence in the technology and its efficacy for the customer's particular applications (about which the customer may, indeed, know more than any peer reviewer). In sum, customer satisfaction cannot substitute for peer review, nor can peer review substitute for customer satisfaction. These related but distinct elements are both important to the program's success.

PEER REVIEW PROCESS

The following sections address the state of OST's peer review program as of April 1998 in terms of the five general steps of a peer review process. It should be noted that in the committee's view, OST has made significant strides in improving its program since the publication of the committee's interim report. Although OST has made many policy changes in response to the recommendations in the interim report, it has not yet implemented all the changes.

Selection of Projects for Peer Review

Forty-three technologies were peer reviewed by the OST program between October 1996 and April 1998. Table 5.1 indicates at what gate in OST's technology development process these peer reviews took place, the number of reviews conducted in FY97, the number completed in FY98 as of May 1, 1998, the number of reviews planned for FY98 (by gate), and the total number of projects that met OST's selection criteria. Because of the relatively low rate of peer reviews conducted under this program relative to the large number of active projects, OST currently funds a large number of technologies that never have been peer reviewed. Many of these projects are in the later stages of development.

TABLE 5.1 Status of Peer Reviewed Technologies in the Stage-Gate Model

	Gate[a]							Total
	1	2	3	4	5	6	Unassigned	
Peer reviews completed FY97	0	7	1	5	8	7	0	28
Peer reviews completed FY98[b]	0	0	1	8	5	1	0	15
Total peer reviews Scheduled FY98[b]	0	0	8	14	8	5	3	38
Projects meeting OST's selection criteria[b]	0	1	10	61	107	3	0	182

[a] The last gate passed by the technology project.
[b] As of May 1, 1998.

Prioritizing Projects for Review

OST's decision to require peer reviews before a technology passes Gate 4 is based on the large increase in funding that typically occurs when a project moves from bench scale (Stage 4 or lower) to field scale (Stage 5 or higher). This is a rational basis for prioritization, and Gate 4 is an appropriate time to schedule a review for projects that have not yet reached field scale. Such a practice, however, does not constitute a sufficient and systematic approach for selecting and prioritizing the projects to be reviewed. In its interim report, the committee recommended that OST develop a rigorous process for selecting projects to be peer reviewed. Such a process should employ well-defined project selection criteria, and OST peer review program staff should be directly involved in making decisions regarding which projects should be reviewed. The committee recommended that OST subject all technologies to a peer review when they enter the program (i.e., at the proposal stage), followed by additional peer review at other critical points in the technology development process (such as at Gate 4). To address the large number of projects at late stages of development that have never been peer reviewed, the committee recommended that in the short term, peer review efforts should focus on high-budget, late-stage projects that have never been peer reviewed but still face upcoming decisions with major programmatic and/or funding implications.

In response to these recommendations, OST has recently issued Version 1.0 of a new policy (DOE, 1998b) describing the three criteria that "must be used in selecting the technologies as candidates subject to peer review" (p. 11), which were used in the selection of projects for FY98 (see Chapter 4). Although OST's three selection criteria are reasonable and should help it choose projects to be reviewed, they do not explicitly address two issues that have been emphasized recently by OST management: (1) deployment of new technologies in the field and (2) the need to reduce funding levels due to budget cuts.

The committee believes that the procedure developed for selecting projects for peer review should ensure that OST would peer review any technology being considered for deployment if it has not already been peer reviewed. Although the suitability of a given technology for deployment at a specific site involves other considerations in addition to the technical merit of the process (e.g., site conditions, infrastructure), technologies should not be deployed without being peer reviewed at some point during development. This applies to all technologies that OST aids in deployment, whether in response to requests from a DOE site, through the Accelerated Site Technology Deployment (ASTD) program (originally named the Technology Deployment Initiative), or though OST's own initiative. Thus, it also applies to new technologies considered for deployment by OST that were developed without OST assistance. **To address these issues, the committee recommends that OST adopt two additional criteria to choose from among those projects that satisfy one of the three existing selection criteria: (1) technologies that are being considered for deployment, and (2) technologies for which a request for further funding has been received or is anticipated.**

The usefulness of peer reviews for projects that are already mature, that is, technologies that have already undergone field trials or initial deployment, has been questioned by several OST personnel during the committee's discussions. The committee notes that reviews of projects just prior to deployment, or even after initial deployment if an additional deployment is planned, could benefit OST and the site at which the technology is to be deployed in several ways. First they provide confirmation as to the validity of the technical process and its probable capabilities and limitations. Second, by providing a basis for comparison of competing technologies, they provide a rationale for technology selection. Third, a peer-reviewed evaluation of technology performance provides the basis for the most defensible cost estimates and hence the most reliable basis for planning future deployment.

Timing of Reviews

The optimal time for a review relative to initial field demonstration will depend on several factors including the relative cost of a demonstration and whether the technology was developed by OST or not. Projects with relatively small dollar value may not warrant a high enough priority to qualify for a Type I or Type II peer review. Reviews of small projects may be handled by mail (Type III review) or by inclusion in a Type I review convened for another related technology. If a program has numerous small projects that are difficult to fit into the Type I or Type II peer review structure, a peer review of all projects in a program could be used. In such a review the panel examines similar projects from a given program or subprogram. However, the range of expertise of the panel would have to include all the relevant areas from the entire suite of projects to be reviewed.

A related problem arises when a technology is developed and deployed without OST involvement. Although application of the current policy of peer reviewing new starts and all projects at Gate 4 will ensure that all technologies developed by OST in the future are appropriately peer reviewed, technologies developed outside DOE and proposed for demonstration by OST would not have been subjected to an OST peer review during development. If these technologies have been deployed successfully at non-DOE sites, they have already undergone field trials. If the cost of demonstration of such projects is relatively low, the most appropriate time for review may be after completion of the initial DOE demonstration. A large group of small demonstrations could be reviewed by a single review panel.

Selection of projects to be included in a Type I review group thus requires consideration of the current state of development of each project, the current status of funding decisions for each project, and the technologies being considered for deployment at the various DOE sites. Because of the broad knowledge required for project prioritization, the direct participation of an OST headquarters program manager in the selection process may be required.

Another important consideration for selecting projects for review is that the technology projects must be at a suitable stage. To evaluate any technology, basic data on system performance are necessary. For example, in a recent Type I review, two projects that had already received funding for construction of pilot plants were reviewed before the pilot plants had operated.[1] Because the pilot plants had not operated, there were no data on which to judge the effectiveness of each process. In this case, it would have been better to postpone review of the projects until after the plants had operated. Because the level of development of these projects was inappropriate for peer review, much of the potential value of these particular peer reviews was lost. The results of the reviews were not, and could not be, used in program management.

Grouping Projects in a Single Review

Although the two additional selection criteria recommended by the committee would assist OST in identifying those projects for which peer review is of highest priority, unless some other changes are made to the peer review program (e.g., increased funding, changes in process), application of these criteria by itself would still leave a large number of projects that are not peer reviewed. **To address this issue, the committee recommends that OST expand its practice of evaluating a number of related technologies in a single peer review, whenever possible.** Such an approach would provide both efficiency in review, by allowing multiple projects to be reviewed by the same panel, and the opportunity for a rigorous comparative evaluation of competing technologies. Thus, if one project is selected for review, related projects could be reviewed concurrently for a marginal increase in cost. Similarity may be determined in at least two ways: by technology or by application. If OST decided to review various technical approaches to a group of similar problems (e.g., for one of OST's focus areas), the review would have to be constructed for this end. Another approach for addressing this issue is discussed in Chapter 6.

Definition of Peer Review Objectives and Selection of Review Criteria

Peer Review Objectives

The committee noted in its interim report that a statement of the objective of the peer review should be made available to all participants in a peer review (i.e., PIs, the ASME Peer Review Committee, all reviewers, and observers) prior to the review so that they clearly understand its context. These

---

[1] At the time of the peer review, one pilot plant was under construction, and construction (but not the permitting process) of the other had been completed.

statements of objectives should be consistent with the general purpose of peer review (e.g., technical, useful for decision making) and should be approved (or revised) by the Deputy Assistant Secretary for Science and Technology or his or her designee.

OST's newly revised Implementation Guidance (DOE, 1998b, p. 13) requires FA/CC program managers, coordinating with the EM-53 Office Manager to submit to the Peer Review Coordinator the objective of the review, technology-specific review criteria, and a list of PIs who will be responsible for providing technical documentation and delivering the technical presentation. Also, although no FY98 peer review report as of April 1998 had included a statement of objectives, OST's Implementation Guidance lists such a statement as a typical feature of the reports. **The committee concludes that OST has developed an appropriate policy but has not yet implemented it. The committee therefore encourages OST to follow up on this new requirement.**

Peer Review Criteria

The committee previously observed that OST's original list of general criteria for peer review included broad issues such as cost-effectiveness, reduced risk, regulatory acceptability, and public acceptance, which include many nontechnical considerations. Certainly it is important to include business criteria and other nontechnical criteria in the management process for deciding which projects or programs to support. For example, such factors as acceptance in the marketplace, political feasibility, and social equity often are just as important in decision making as technical concerns. However, if a technology does not perform up to technical requirements, the degree of customer satisfaction, cost, and feasibility are, in a sense, beside the point. Nevertheless, the peer review processes that the committee defines and discusses here focus exclusively on technical criteria.

In other types of reviews (such as business reviews) it may be useful and even necessary to integrate business and technical criteria. Such broad-scope program reviews may well include not only technical criteria related to the specifications and functionality of a technology but also such issues as payback period, implementation costs, and spillover effects of a technology (e.g., economic development impacts). In a peer review, however, such broad, mostly (or partially) nontechnical issues may sidetrack reviewers into issues that are outside their expertise or are difficult to resolve within the time constraints of a two- to three-day review. For example, if technologies are peer reviewed very early in their development, both regulatory and public acceptability may be indeterminable. Thus, in peer review, where technical merit is the focus, business and technical criteria generally should not be integrated.

The committee therefore recommended in its interim report that OST revise these broad criteria to focus on technical aspects of the issues mentioned above, or remove them from the list of review criteria. The committee also recommended that OST develop a well-defined general set of technical criteria for peer reviews, to be augmented by technology-specific criteria as needed for particular reviews. These technology-specific criteria should supplement, not supplant, the basic criteria: they should not deflect the review from completing the basic evaluation.

The general review criteria released by OST in early 1998 (DOE, 1998b) and listed in Chapter 2 are consistent with the committee's recommendations. If these criteria are implemented wholeheartedly they should result in a major improvement in OST's peer review program. OST's new policy also states that these general review criteria will be augmented by technology-specific criteria, as recommended in the interim report. As mentioned in Chapter 4, ASME combines the general and technology-specific review criteria to develop the specific criteria for each peer review, thus enabling reviewers to focus on the specific objectives of a given review. In addition, even though these review criteria form the basis of the review panel's evaluation, the review panel has the authority to pursue other issues that arise (DOE, 1998b). **The committee finds that these revised general criteria and the procedure for developing technology-specific criteria are a meaningful improvement over the original review criteria because they allow OST's peer review program staff to focus the reviews on important technical issues. The procedure also has sufficient flexibility to allow the review criteria to vary as a function of the stage of development of a technology.**

Selection of Peer Reviewers

Qualifications

OST's criteria for reviewer selection are discussed in Chapter 4. Although the committee finds that OST's selection criteria are generally adequate to ensure the technical credibility of the review panel, in its interim report the committee recommended that OST consider modifying the criteria to emphasize expertise relevant to the review. For example, its first criterion (education and relevant experience) could specifically include knowledge of the state of the art of an aspect of the subject matter under review, including national and international perspectives on the issue.

Panel Formation

Another important factor in selecting a peer review panel is that all relevant areas of expertise required to address the review criteria should be represented on the peer review panel. This is particularly important for the very diverse and complex technologies that are being developed by OST for environmental cleanup of the DOE complex, which may require large review panels. In its interim report, the committee recommended that the scope of the review panel, and thus its size, should depend on a number of factors, including the complexity and number of projects being reviewed and the specific review criteria established for the peer review. The committee also suggested that OST use a large database of potential reviewers to help identify those with relevant expertise. The databases currently used by RSI to identify committee members are described in Chapter 4.

Although the range of expertise on peer review panels observed by this committee has been acceptable, the committee believes that the databases may not have adequate scope to identify the broad range of reviewers likely to be necessary for some complex projects. In its interim report the committee noted that it would be appropriate for OST to gain access to databases from other professional societies or review organizations. ASME's *Manual for Peer Review* (ASME, 1998a) states that ASME will work to develop such collegiality agreements in situations where expertise beyond that available to ASME is required, but none have been employed to date[2]. To the committee's knowledge, OST has not addressed this issue. **The committee therefore recommends that OST establish a more systematic approach for accessing reviewer information from other databases (e.g., chemical engineers, geologists, physicists, materials scientists, biologists) and from other professional societies, as needed to ensure the appropriate range of expertise for all review panels.**

---

[2]During the final stages of report publication, the committee learned that ASME has in fact established agreements with the American Chemical Society, the Society of Environmental Toxicology and Chemistry, and the Federation of American Societies for Experimental Biology to assist in the identification of qualified reviewers.

Conflict of Interest

In its interim report, the committee recommended that in order to ensure the independence of peer reviewers, OST should include in its requirements a criterion that explicitly excludes EM staff and contractors with real or potential conflicts of interest, including all OST staff and contractors, from consideration as peer reviewers. OST has revised its conflict-of-interest policy to explicitly exclude "all DOE staff and contractors with real or potential conflicts" from consideration as peer reviewers (DOE, 1998, p. 7); *in practice*, OST has interpreted this policy to exclude all DOE staff, whether or not there is a real or potential conflict of interest with the projects under review. This interpretation goes beyond the committee's recommendation but is consistent with the ASME conflict-of-interest policy.

The committee would like to point out that concerns over conflicts of interest should not necessarily preclude all DOE staff and contractors from serving on a peer review panel, however. The committee believes that DOE staff from organizations outside OST and its contractors could be used in special circumstances when the appropriate expertise is not available outside DOE (e.g., high-level waste tanks) and where these individuals have had no connection with the projects under review. For some reviews, for example, national laboratory researchers may be unique sources of expertise for technologies addressing certain DOE cleanup problems. In such cases, issues of potential conflicts of interest would have to be treated very carefully and openly. The reviewer selection process should in general avoid DOE staff as peer reviewers, however, and should ensure that the DOE-affiliated persons are never more than a small fraction of a panel's membership.

Planning and Conducting the Peer Review

Organization and preparation of each peer review event are key to a successful peer review program. Well in advance of the review, peer reviewers should receive written documentation on the significance of the project and a focused charge that addresses technical review criteria. When a review panel is convened, it should be provided with clear presentations by the project team, as well as adequate time to assess the project comprehensively so that the panel is able to write a report that effectively summarizes and supports its conclusions and recommendations. Issues such as proprietary information also have to be handled in a systematic and confidential manner.

Documentation

The documentation required for an OST peer review is presented in Chapter 4. This list has been modified from OST's original list of required documentation in response to the committee's interim report, which pointed out that some of the required documents addressed nontechnical issues. The revised list identifies some of the documents required to evaluate the technical merit of a project and, if implemented, should improve the quality of background materials provided to peer review panels. One document that is not included in this revised list of required materials is a statement of work, or proposal, that describes the specific activities that will be carried out if the project is funded. **The committee recommends that a detailed proposal or statement of work be required for all peer reviews.**

Confidentiality of Technical Information

During some early OST peer reviews, members of the committee observed problems with how confidential technical information was handled. In particular, confidential materials that were necessary to evaluate a project were withheld from the review panel, which prevented panelists from judging the technical merit. The confidentiality of proprietary technical material has been ensured in more recent reviews, such as the review of the Radioactive Isolation Consortium's SMILE and SIPS technologies, conducted on July 8-9, 1997, in Columbia, Maryland, and the review of the Industry Program on January 21-22, 1998, in Columbia, Maryland. Unfortunately, at the latter, procedures restricting the panel's review of proprietary materials to the site of the review unnecessarily restricted the time reviewers had to digest these proprietary materials. In future cases where proprietary information cannot be sent to the panels prior to review, OST may want to consider convening the panels at the peer review site one day prior to the review to ensure sufficient time to review this material. In general, however, the system for handling proprietary information appears to be appropriate and effective.

Anonymous Versus "Open" Peer Reviews

As discussed in Chapter 2, another important consideration in planning and conducting peer reviews is whether the evaluations should be conducted anonymously or openly (i.e., using publicly known reviewers). In the case of OST, the committee believes that the strengths of open reviews (e.g., enhanced credibility of the process, the potential for more constructive evaluations) far outweigh the potential weaknesses (e.g., possible lack of candor by some reviewers when evaluating weak proposals), especially for the peer review of projects or programs.[3] As such, the committee encourages OST to continue to promote openness of its peer reviews and to fully inform the public and others attending the reviews of their nature. An approach used by the EPA Science Advisory Board is to make lists of the reviewers and their affiliations available and to have each reviewer publicly state his or her pertinent experience and any factors that could affect bias at the beginning of the peer review. Similarly, at the OST peer review of Industry Programs held in January 1998, members of the review panel were asked to give short descriptions of their backgrounds. The committee encourages OST to continue this practice in future reviews. In addition, at the beginning of each review, the chair of the peer review panel should clearly explain the objectives of the review, the specific review criteria that will be addressed by the panel, and how the results of the peer review will be used in OST's decision-making process.

Usefulness of Peer Review Results

Peer Review Reports

OST's procedures for peer review reporting are described in Chapter 4. Although some of the early peer review reports did not document the reasoning for conclusions or provide adequate support for recommendations, recent reports have more clearly explained the rationale for the panel's conclusions and recommendations. Two other improvements in peer review reports have been the addition of a section that summarizes the review criteria and the inclusion of short biographical sketches of the peer reviewers. In its interim report, the committee recommended that the peer review reports could be improved further

---

[3] If OST decides to implement a system of peer review to evaluate large numbers of research proposals, however, it may want to consider the pros and cons of anonymous versus open peer reviews for this purpose.

by also including a statement of the objective of the review and a list of references used in the analysis. These additions could improve the overall quality of the reports by increasing the reviewers' focus on their charge and could make the reports more useful to program management by more clearly documenting the basis of the review.

In response to this recommendation, OST recently established a new policy that requests ASME peer review reports to include the features of technical peer review reports listed in Chapter 4 (DOE, 1998b), including the objective of the review and a reference list. As of April 1998, neither of these items had yet appeared in the peer review reports, and the committee recommends that OST implement this new policy.

Feedback Procedures

OST's procedures that governing the feedback of peer review results into program management and development decisions are described in Chapter 4. The committee has reviewed OST's written procedures on feedback and finds that they are reasonable, although the committee has observed that the implementation of many of these procedures has been uneven. Many peer reviews are not linked to decision points, despite previous and current policy guidance.

The committee noted in its interim report that the timeliness and quality of the formal written responses from the FA/CC program managers also have been uneven. None of the formal responses from the program managers for the FY97 reviews were transmitted to ASME within 30 days of the peer review, as required by OST policy.[4] In addition, many of the written responses did not document how OST intended to follow through with the conclusions and recommendations of its peer review reports. For example, although the review panel for the Cost of In Situ Air Stripping of VOC Contamination in Soils Project made 10 specific recommendations for improving the project, OST's entire formal response was, "The Program Manager agreed with the recommendations of the Review Panel, and the report is being revised" (ASME, 1997, p. 41). Such responses are not useful to OST decision makers charged with funding decisions.

To address such problems, in its interim report the committee encouraged OST to enforce the 30-day requirement for written responses and to require more detailed responses that fully describe how the recommendations of

---

[4]Personal correspondence from Sorin Straja, RSI, to Erika Williams, NRC, March 26, 1998.

the peer review reports were implemented or considered. OST has indicated its intent to implement these recommendations; for example, the Implementation Guidance states that a schedule for acting on recommendations must be submitted to the EM-53 Office Director within 30 days of receipt of the Technical Report (DOE, 1998b, p. 7). The committee notes that since the release of the interim report, both the quality and the timeliness (although still none have been submitted on time) of the written responses from OST program managers have increased significantly. The committee notes that the need for a formal response to reviewer comments from FA/CC program managers (and the inability to meet the 30-day deadline) is a strong indication that peer review has not yet become a part of the "culture" of OST, whereby FA/CC program managers would embrace peer review as a valuable tool to help guide their decision process (see discussion in Chapter 7).

6

"Triage" Approach for Reducing Project Backlogs

OST has participated in more than 800 environmental restoration technology development projects since 1989 (DOE, 1998b), 226 of which are currently funded technologies and most of which have not been peer reviewed. Peer review of these technologies would be invaluable in helping determine which are ready for deployment,[1] which merit further development, and which should be canceled. The number of technologies is too large to be reviewed rapidly by the current OST peer review process, however, because only 28 peer reviews were completed in 1997 and only 38 reviews have been scheduled for FY98 (Table 4.1).[2] The relatively small percentage of active projects that have been peer reviewed also raises questions about the effectiveness of the peer review process, because it does not allow a consistent application of peer review results in OST's decision-making process. For example, projects that have been peer reviewed may be more likely, or less likely, to be funded solely because they were reviewed, rather than because of their technical merit.

In Chapter 4, the committee recommends two procedural changes to improve both the rigor of the selection process and the efficiency of the review process. Although these changes would help deal with OST's backlog of projects, eliminating the problem would require more fundamental changes in its peer review process. In this chapter, the committee focuses specifically on how OST could reduce this large backlog of OST-supported projects that have never been peer reviewed.

OST's current practice in which nearly all peer reviews include formal presentations by the project team, followed by deliberations by the panel, and further question-and-answer sessions over the course of two to three days, places

---

[1] In this context, deployment refers to large-scale application of a process.
[2] As of April 15, 1998.

significant limits on the number of projects that can be peer reviewed by a single panel.³ Even if the number of projects that were peer reviewed during a single Type I review could be increased by improved efficiency, OST's backlog of technologies that have never been peer reviewed still would take years to be eliminated through its current process. If OST is to fulfill its policy that "all projects are to be peer reviewed" in the short-term (i.e., the next year), it will have to make significant changes in how peer reviews are conducted.

The committee recommends that OST consider adopting a "triage" approach that would allow far greater numbers of technologies to be peer reviewed. This approach would involve a formal prescreening of projects by peer reviewers based exclusively on the written documentation on the project— in effect, a "mail review" of projects, to be followed by a formal meeting of the panel to discuss and rank the projects. During this prescreening review, panel members would be asked to rank all related technologies in a given area that are being considered for additional development or deployment.⁴ Rankings from the panel as a whole could then be used by OST program managers to determine those highly ranked, low-budget projects that should be considered for funding without additional peer review; those highly-ranked projects that should receive a more detailed evaluation (including presentations by the project team and question and answer sessions); and those technically weak projects that should not be considered for funding. This approach would provide OST program managers the basis for discontinuing funding for technically weak projects and might provide them with sufficient technical input (to be supplemented by input on nontechnical factors) to make a decision to fund a low-budget project. The prescreening evaluations should not be used as the sole means for providing technical input into decisions to fund high-budget, environmental remediation projects, however. Peer reviews involving presentations by the project team and question-and-answer sessions should be carried out for all projects involving significant capital investment by OST.

Because prescreening evaluations require only written documentation on the projects to be reviewed, the triage approach could be used to evaluate a large number of projects at a single review. For example, a single review could evaluate all projects developed to address a specific type of environmental management problem or an entire OST FA/CC program. The approach also could include related projects from OST's inventory of nearly 600 projects that

---

³ According to RSI staff, the maximum number of projects that can be peer reviewed effectively during a Type I review using the current OST process is approximately five.

⁴ If the prescreening review involves technologies at very different stages of development, it might be necessary to develop somewhat different review criteria for each general stage of development. The same reviewers could conduct all of the reviews, however.

are not currently funded. This might allow OST to identify especially promising technologies within its inventory that should be funded for demonstration or deployment.

Because of the breadth of technical issues that might be evaluated during a review of a large number of technologies, the number of reviewers required for the prescreening evaluation could have to be significantly larger than the five to six members typically employed for an OST Type I review. If the prescreening review process results in a greatly reduced number of technologies requiring a formal, panel review (as intended), fewer reviewers might be needed for the panel review itself, however.

Increasing the *efficiency* of OST's review process through the triage approach outlined above could result in some trade-offs in terms of the *quality* of the peer reviews, however. The same level of detail and thoroughness that is possible through panel reviews involving lengthy interaction and questioning of the project team would not be possible in the prescreening evaluations. The difference in quality of the reviews could be minimized by ensuring that the written documentation provided to the panel is complete and of high quality. A detailed proposal or statement of work for every project, as recommended earlier in this report, would be especially important for these prescreening reviews. The committee believes that the potential decrease in quality of some peer reviews, however, would be more than offset by the increased effectiveness of a peer review system in which all projects funded by OST have been through a credible peer review.

7

Improving the Effectiveness of OST's Peer Review Program

The committee is encouraged that OST continues to actualize its new goal of implementing peer review of its activities. As stated earlier in this report, the committee finds that the foundation of the peer review program is sound but the process is still in the early stages of development or "maturity." **Despite the marked improvements in the *procedures* for conducting peer reviews over the past year, OST's peer review program still has not fully achieved its stated objectives of providing high-quality technical input *to assist in decision making*.** The first step for OST leadership is to ensure that peer reviews are effectively linked to OST decision making. In order to continue to develop and achieve a more effective peer review program, OST leadership also will have to commit to a process of continuous assessment and improvement involving cycles of planning, execution, and evaluation. This will require incorporating process improvement as part of daily activities, identifying and eliminating problems at their source, and striving toward improvement by implementing solutions to problems as they are identified. The basis for evaluating the *efficiency* and *effectiveness* (i.e., the use of peer-review results in better management decisions and in program improvement) of the peer review program should be as quantitative as possible, based on fact, and oriented toward obtaining results. In this chapter, the committee outlines a number of approaches that OST leadership should consider as it works to improve the effectiveness of the peer review program. The committee also discusses some of the institutional factors that will have to be addressed to implement a credible, efficient, and effective peer review program.

LINKAGE OF PEER REVIEWS TO MANAGEMENT DECISIONS

One of the most significant problems with the current peer review program is that many peer reviews are not linked to decision points, in spite of previous and current policy guidance. The committee stresses that a peer review report finding a project to be technically sound, based on good science, and capable of practical realization should be a necessary but not sufficient condition for passing certain TIDM gates. A project could fail to pass the gate for programmatic reasons even if it were technically sound, but a report stating that a project is not technically sound should be a sufficient reason to reject a project at any gate. The project need not necessarily be terminated (e.g., it might need more development and re-review before moving to the next stage, or the review might be "appealed" and reconsidered, especially if the panel was divided in its conclusions), but it should not pass a gate in the face of an adverse peer review.

To have any effect on programmatic decision making, peer reviews must occur well before project decisions are made. The timeliness of reviews has been one of the most significant problems with the program. A number of FY97 peer reviews were completed after project decisions had been made, and in these instances, few benefits of peer review were realized. For example, the decision to fund the next stage of development of the In Situ Redox Manipulation project had been made prior to peer review of the project. In another peer review (the Large-Scale Demonstration project at Fernald Plant I), a major portion of the facility had already been decommissioned when the peer review was conducted. Although retrospective reviews can provide some guidance to other projects, if OST's new peer review program is to be effective, peer reviews must occur prior to key points in the technology development decision process. Because for FY98, the FA/CC program managers selected the projects to be reviewed, the times for review, and the technology-specific review criteria, the committee expects that the results of the peer review should fit more logically with the decision-making process. The ASME Peer Review Committee also has identified the lack of a clear relationship between the peer review results and OST decision making as a recurring problem with the peer review program.[1]

To address this issue, the committee recommended in its interim report that OST develop a targeted plan for the peer review program. The plan should consider factors such as how many of OST's technology projects can be peer reviewed, realistic schedules for the reviews, and the peer review program budget. To be effective, this plan also should ensure that peer reviews are conducted early enough in the budget cycle to allow their results to be used as an input into meaningful funding decisions. In developing its plan for the peer

---

[1] Discussions at ASME's PRC meeting, January 26, 1998.

review program, OST should consider expanding its practice of consolidating reviews of related projects into a single review (i.e., a Type I review) or several overlapping reviews in order to increase the number of projects that can be reviewed with the resources available. Another value of reviewing multiple projects during a single peer review is that it tends to ensure that the projects reviewed together are judged by the same standard; thus, it normalizes the results (Kostoff, 1997b). To the committee's knowledge, a targeted plan has not yet been developed, but the Implementation Guidance gives the Peer Review Coordinator responsibility for developing such a plan (DOE, 1998, p. 19).

The linkage between peer review results and OST's decision-making process also could be improved by explicitly identifying where and how the results of peer reviews will be used, *before* the review is conducted. Therefore, **the committee recommends that as part of the documentation provided to peer review program management during the process of selecting projects for review, OST program managers be required to clearly identify the upcoming decision or milestone for which the results of the peer review will be used.** This information also should be provided to peer reviewers as part of the documentation that they receive in preparation for the review

EVALUATION AND IMPROVEMENT MECHANISMS

Benchmarking

One approach for guiding the development of an internal evaluation procedure for the peer review program would be for OST peer review program managers to proactively seek out and learn from other organizations that have more mature peer review processes. This process of learning from the practices of other organizations is called "benchmarking." Benchmarking is a process-oriented, systematic investigation in which an organization measures its performance against that of the "best in class" (i.e., other organizations with renowned peer review programs) to determine what should be improved. Benchmarking involves searching for new ideas, methods, practices, and processes; adopting the practices or adapting the good features to the specific needs of the organization; and implementing them to improve the effectiveness of the program.

There are usually four elements to the benchmarking process:

1. Planning: identify internal targets for benchmarking (specific opportunities for improvement),
2. Analysis and preparation: understand the current process, form an evaluation team, and identify the external organizations whose processes constitute the benchmark,
3. Integration: understand the process in the external organization and establish new process goals (prioritize, plan and test proposed solutions), and
4. Implementation: develop action plans, implement changes, monitor performance, and recalibrate the benchmark.

Benchmarking enables learning from the leadership and experience of others. It should challenge current internal paradigms of process performance, provide an understanding of opportunities and methods for improvement, and identify strengths within the organization. It also should result in establishing goals driven by results and in providing insights that will assist in prioritizing and allocating internal resources. One strength of benchmarking is that it provides options and ideas with proven performance, rather than relying on "new" ideas developed from within the organization. It is a proactive method for developing measures of effectiveness because it is based on objective evaluation rather than "gut feel" or perception.

Peer review programs in some organizations that could be used by OST in such a benchmarking process are described in the boxes throughout Chapter 3. Based on its analysis of the development of OST's peer review process since its inception in October 1996, the committee believes that it would have been extremely beneficial if OST had used such a benchmarking process to help design its new peer review program before it was established.

To encourage the development of a more effective peer review program, OST management should support benchmarking by focusing on the processes that are critical to the peer review program, by being open to new ideas, by being willing to admit that its current process is not the best, and by committing to provide resources for change and to overcome resistance to change. The benefits of improving the peer review process through benchmarking must be weighed against the possible negative effects of constantly changing procedures, however. Benchmarking efforts should be targeted to specific weaknesses in the peer review process and should be initiated at a logical time in OST's annual peer review cycle when specific areas of improvement have been identified (e.g., after the release of this report, or after the annual ASME Peer Review Committee meeting), that is, when the first two steps of the benchmarking process (i.e., planning and analysis) already have been completed. The overall goal of the evaluation and improvement process should be a high-quality, relatively stable peer review process. The benchmarking process also should involve the development of metrics to quantify the efficiency and effectiveness of the peer review program.

Metrics

A world-class peer review program has to be built on a foundation of measurement, data, and analysis. Measurements should be directly related to the goals, or objectives, of the program. Metrics are defined as performance indicators that are measurable characteristics of services, processes, and operations used by the organization to track and improve performance. Useful metrics share a number of important characteristics: (1) they encompass the key outputs and results of the process steps, such as performance, program impact, and cost; (2) they are based on systematically collected data rather than anecdotal observations; and (3) they are selected to represent factors that lead to the best customer, operational, and financial performance. A system of metrics tied to organizational performance requirements represents a clear and objective basis for aligning activities with the organization's goals. It is important to note that the metrics themselves can be evaluated and changed when analysis of data from the tracking process suggests they should be changed. Benchmarking against other peer review programs (see previous section) can be used to calibrate metrics as above or below the norm of performance.

Metrics can be used to assist in the measurement of effectiveness and can help evaluate the success of a program in realizing its objectives. Two types of metrics can be considered: activity metrics and performance metrics. Activity metrics are an indication of the *efficiency* of the process, whereas performance metrics are an indication of the *effectiveness* of the program, (i.e., achieving the desired results). Although OST has not yet established metrics or a benchmarking process for its *peer review* program, it has begun to develop performance metrics for its technology programs as part of its annual performance planning. The DOE Environmental Management Advisory Board's Technology Development and Transfer Committee also recently held a workshop to develop guidelines for formulating performance measures for research, technology development, and deployment. This workshop served as a form of benchmarking by involving panelists from other federal agencies with well-established systems of performance measures.

Properly chosen and clearly defined metrics could be a powerful management tool to help OST improve the efficiency and effectiveness of this program. Activity metrics that could be chosen include the following:

- the percentage of projects reviewed at each gate,
- the percentage of reviewed projects that were not funded in the next gate review decision,
- the percentage of adequate DOE written responses to peer reviews that were received within the required 30 days,
- the degree of follow-up to recommendations of peer review panels,
- the number of peer reviewed articles published on OST-funded technologies, and
- the percentage of peer reviews conducted in which satisfactory program documentation was provided to the peer review panel at least two weeks prior to the review.

Because OST is at the beginning stages of the peer review process, it is not clear if the current activity level would yield a statistically significant measurement; however, over time, the data should become significant.

Performance metrics should be based on OST's success criteria. Proper use of well chosen and articulated metrics can result in decisions that improve the strategic alignment of the project with the needs of customers or end-users. Paladino and Longsworth (1995) provide a broad description of proposed decision gate criteria for the TIDM, and OST recently has provided an updated list of review criteria for the peer review program (DOE, 1998b). OST also should clarify what would be successful performance relative to the criteria, however. For example, project impact and the percentage of resources going to successful projects could be appropriate performance metrics. Other metrics could include

- the number of recommendations in the peer review report that have been implemented;
- the number of decisions affected by the peer review results (i.e., not whether the decision is positive or negative, but whether the decision is affected by the peer review report);
- the amount of funding reallocated as a result of the report; and
- the number of changes in research paths (midcourse corrections) that occur as a result of peer reviews.

These activity and performance metrics are provided as examples— ultimately OST management will have to establish its own set of metrics based on the success criteria it sets for the technology development program. **The committee recommends that OST management develop an effective evaluation and improvement process for the peer review program that** includes regular benchmarking against other peer review programs and the collection of activity and performance metrics.

POTENTIAL APPLICATIONS OF PEER REVIEW WITHIN OST

Because OST has chosen to focus its new peer review program on the review of individual projects at various stages of development, the committee also has focused this report on the peer review of projects. As discussed in Chapter 3, however, the fundamental principles of peer review (i.e., independent, external, expert, technical) also can be applied to programs, subsets of programs, or technical needs. One potential application of these types of peer reviews within OST would be to evaluate the research and development efforts that are needed to address the environmental problems at contaminated sites. Another potential application within OST would be to assess the technical balance of OST programs in the context of other programs, both within DOE (and OST in particular) and outside the DOE complex.

Although these and other applications of peer review within OST are possible, simply adding to the number of reviews will not solve OST's problems. OST already has an extensive system of internal and external reviews that it uses to assist staff in making programmatic decisions (see Appendix B). The development of a credible peer review system to evaluate the technical merit of proposals, individual projects, and programs, however, might allow OST to discontinue some of these other types of reviews, if they serve the same or similar objectives. **The committee recommends that OST carefully evaluate the objectives and roles of all of its existing reviews, and then determine the most effective use of peer reviews (of various types) in meeting its overall objectives.**

For any type of peer review that OST chooses to implement, however, the "expert, independent, external, and technical" criteria should be applied to achieve the objectives of the review. That is, the reviewers should have the technical background required to make the judgments called for and they should have no conflict of interest. For peer reviews of programs, the *collective* expertise of reviewers should also be appropriate to projects within the program, and each panel member should have a broad knowledge of the area covered by the range of projects. For specific reviews with clear objectives that involve technical matters such as economics, risk assessment, or other socioeconomic issues, supplementing the technologists with experts in these areas (e.g., alternative dispute resolution) may be appropriate.

If OST decides to implement a peer review process for evaluating proposals, the technical merit and balance of its programs, or its program "needs," such peer reviews need not be arranged under the same ASME program as project reviews. In fact, some reviews might be conducted by a standing group parallel to the Environmental Management Advisory Board (EMAB) but composed of world-class engineers and scientists (another example of such a standing group, the NRC Board on Assessment of NIST has been described in Box 3.5).

OST'S "ORGANIZATIONAL" CULTURE AND LEADERSHIP

One feature characteristic of organizations that effectively use peer review as a tool for management of their research and development portfolios is a peer review process that is ingrained in their organizational culture[2]; in other words, for these organizations, peer review is "standard operating procedure" for providing input to their decision-making process. Even after deciding on a peer review process that seems adequate on paper, OST still will have to change its organizational culture so that it embraces peer review as an essential part of its decision-making process.

The peer review culture is not yet ingrained within some parts of the Department of Energy, especially the EM program (NRC 1995b,c, 1996; GAO, 1996). This may derive from a time when DOE laboratories were working in some fields where the national expertise was predominantly within the DOE organization. Today, programs like those of OST are not unique to DOE but are shared by other organizations, and a broad range of expertise is available outside the DOE "family." OST is just beginning to turn outside for technical advice, however. The benchmarking process described above would reinforce this positive trend. In addition, the committee is encouraged that the EM Science Program, which funds basic environmental research of relevance to EM, has embraced peer review for assessing the scientific merit of proposals.

A change in the organizational culture of OST will require leadership. Kostoff (1997a) recently pointed out that the commitment of an organization's senior management to high-quality reviews is one of the most important factors for high-quality peer review programs. OST management has begun this process of change by prescribing the use of peer review of projects at several stages of development. Although such leadership pressure provides a strong motivating force for the *use* of peer review, it will not by itself result in a change of organizational culture, which must pervade all levels of the organization.

---

[2]Organizational culture can be defined as the normative values of its members.

Corporations that introduce and maintain effective safety and quality assurance programs provide an example of how such a change in value systems can be accomplished. When these organizations adopt such a program, corporate managers typically employ a strategy that involves a steady flow of information to staff at all levels of the organization over a period of time. Such communications include educational materials about the program itself, case histories of how the program addressed specific issues or problems, and specific data on how effective the program was in achieving its objectives (i.e., performance metrics).

In the case of peer review in OST, individual members of the organization will value peer review when they see beneficial results (e.g., which might be disseminated by using case histories), when management gives them logical messages of the value of peer review, and/or when they have incentives to use it or disincentives not to use it (Kostoff, 1997b). **The committee recommends that OST leadership develop an explicit strategy to accomplish a change in its organizational culture by distributing (1) educational materials that summarize the basic principles and benefits of peer review as a tool to decisionmaking, (2) case histories illustrating how peer review input served to improve specific projects, and (3) summaries of key performance metrics that demonstrate how peer reviews are used to meet the overall objectives of OST's program.**

References

Abrams, P. 1991. The predictive ability of peer review of grant proposals: The case of ecology and the U.S. National Science Foundation. Social Studies of Science 21:111-132.

American Institute of Physics. 1997. Sensenbrenner Chides Science Agencies. The American Institute of Physics Bulletin of Science Policy News 103 (August 18).

ASME. 1997. Assessment of technologies supported by the U.S. Department of Energy Office of Science and Technology: Results of the Peer Review for Fiscal Year 1997. American Society of Mechanical Engineers, Center for Research and Technology Development.

ASME. 1998. Manual for Peer Review. American Society of Mechanical Engineers, Center for Research and Technology Development.

Bozeman, B. 1993. Peer review and evaluation of R&D impacts. Chapter 5 in Evaluating R&D Impacts: Methods and Practice, B. Bozeman and J. Melkers, eds. Boston: Kluwer Publishing.

Chubin, D. 1994. Grants peer review in theory and practice. Evaluation Review 18(1):12-19.

Chubin, D., and E. Hackett. 1990. Peerless Science: Peer Review and U.S. Science Policy. Albany, N.Y.: State University of New York Press.

Cole, S. 1991. Consensus and reliability of peer review evaluations. Behavioral and Brain Sciences 14 (1):140-150.

Committee for Economic Development. 1998. America's Basic Research: Prosperity Through Discovery. New York. Committee for Economic Development.

Conway, R.A., W.H. Patrick, Jr., and C.H. Ward. 1996. GPRA Review of the Environmental Laboratory of the U.S. Army Corps of Engineer Waterways Experiment Station. Vicksburg, Miss.

REFERENCES

Conway, R.A., K.L. Dickson, and C.H. Ward. 1997. GPRA Review of the Environmental Laboratory of the U.S. Army Corps of Engineer Waterways Experiment Station. Vicksburg, Miss.

Cooper, R. G. 1993. Winning at New Products, 2nd edition. New York: Addison Wesley Publishing.

Cozzens, S.E. 1987. Expert review in evaluating programs. Science and Public Policy 14(2):64-71.

DOD (U.S. Department of Defense), Strategic Environmental Research and Development Program, Science Advisory Board. 1998. Annual Report to Congress, Fiscal Year 1997. Arlington, Va.

DOE (U.S. Department of Energy). 1996. Draft Description of OST Departmental, Program & Project Level Reviews. Washington, D.C.: U.S. Department of Energy.

DOE. 1997. Technology Decision Process Procedure: Working Copy, Revision 7.0. Washington, D.C.: U.S. Department of Energy.

DOE. 1998a. Accelerating Cleanup: Paths to Closure, Draft. Washington, D.C.: U.S. Department of Energy, Office of Environmental Management.

DOE. 1998b. Implementation Guidance for the Office of Science and Technology Technical Peer Review Process. Version 1.0. Chicago: U.S. Department of Energy, Center for Risk Excellence.

GAO (U.S. General Accounting Office). 1996. Energy Management: Technology Development Program Taking Action to Address Problems. GAO/RCED-96-184. Washington, D.C.: U.S. General Accounting Office.

Institute of Regulatory Science (RSI). 1998. Handbook of Peer Review. Institute of Regulatory Science, Columbia, MD.

Koning, R.N. 1990. Peer review. Scientist 4(17):12-14.

Kostoff, R.N. 1997a. Peer Review: The appropriate GPRA metric for research. Science 277:651-652.

Kostoff, R.N. 1997b. Research program peer review: Principles, practices, protocols (on-line companion paper to Kostoff [1997a], available at http://www.dtic.mil/dtic/kostoff/index.html).

Moxham, H., and J. Anderson. 1992. Peer review: A view from the inside. Science and Technology Policy 5(1):7-15.

NRC (National Research Council). 1995a. Allocating Federal Funds for Science and Technology. Washington, D.C.: National Academy Press.

NRC. 1995b. Committee on Environmental Management Technologies Report for the Period Ending December 31, 1994. Washington, D.C.: National Academy Press.

NRC. 1995c. Improving the Environment. Washington, D.C.: National Academy Press.

NRC. 1996. Environmental Management Technology-Development Program at the Department of Energy: 1995 Review. Washington, D.C.: National Academy Press.

NRC. 1997a. Building an Effective Environmental Management Science Program: Final Assessment. Washington, D.C.: National Academy Press.

NRC. 1997b. Peer Review in Department of Energy–Office of Science and Technology: Interim Report. Washington, D.C.: National Academy Press (available at http://www.nap.edu/readingroom).

NSF (National Science Foundation). 1995. Grant Policy Manual. NSF 95-26. Arlington, Va.:National Science Foundation.

NSF. 1997. Grant Proposal Guide. NSF 98-2. Arlington, Va.: National Science Foundation.

OTA (Office of Technology Assessment). 1991. Federally Funded Research: Decisions for a Decade. OTA-SET-490. Washington, D.C.: Office of Technology Assessment.

Paladino, J., and P. Longsworth. 1995. Maximizing R&D investments in the Department of Energy's environmental cleanup program. Technology Transfer (December):96-107.

Porter, A., and F. Rossini. 1985. Peer review of interdisciplinary research proposals. Science, Technology and Human Values 10(1):33-38.

Royal Society. 1995. Peer Review: An Assessment of Recent Developments. London: The Royal Society.

USNRC (U.S. Nuclear Regulatory Commission). 1988. Peer Review for High-Level Nuclear Waste Repositories: Generic Technical Position, by W.D. Altman, J.P. Donnelly, and J.E. Kennedy. NUREG-1297. Washington, D.C.: U.S. Nuclear Regulatory Commission.

Appendix A

Description of OST's Technology Investment Decision Model[1]

OST's Technology Investment Decision Model is a procedure OST developed to provide a common basis on which to assess and manage the performance, expectations, and transition of technologies through the development process (Paladino and Longsworth, 1995). It is a user-oriented decision-making process for managing technology development and linking technology development activities with cleanup operations. It should be noted that the TIDM procedure has been documented by OST, but although OST does use the overall TIDM framework to track its projects, OST as a whole has not yet adopted this approach in its decision making.

The TIDM identifies six R&D stages from basic research[2] through implementation of a technology (see Figure 4.1). At each stage, specific criteria, requirements, and deliverables form a common basis for technology assessment. In the model, stages are separated by "gates"—decision points at which projects are evaluated for funding of the next stage. The "stage-gate" process is meant to provide for evaluation of projects at all stages of development against technical and nontechnical criteria selected to ensure that the technologies developed will

---

[1] The material in this appendix is based on OST's descriptions of its procedures, not on the committee's evaluation.

[2] Since 1996, OST has funded "mission-directed" basic research through the Environmental Management Science Program (EMSP). Unlike many basic research programs, EMSP is explicitly focused on DOE's environmental management problems. The specific objective of this program is to improve the long-term effectiveness of DOE's cleanup effort by involving basic researchers from universities, national laboratories, and the private sector in long-term research to address DOE's most challenging environmental management problems.

provide superior performance, will meet the acceptance requirements of the intended customers, and can be moved into the marketplace.

At each gate of the TIDM, OST's FA/CC program managers are responsible for evaluating a technology's documentation in accordance with the appropriate criteria. Programmatic driver criteria to enter each stage include technology end user need; technical merit; cost; safety, health, environmental protection, and risk; stakeholder, regulatory, and tribal issues; and commercial viability. If the FA/CC program manager determines that the technology warrants passing through a gate, the technology maturation process will continue. If the program manager determines that the technology does not warrant further consideration, funding is discontinued. The six stages (and gates) and a description of each, including goals, objectives, and measures of effectiveness, follow.

The following material was adapted from DOE's *Technology Decision Process Procedure* (DOE, 1997); and from "Maximizing R&D Investments in the Department of Energy's Environmental Cleanup Program," by Paladino and Longsworth (1995). See also Figure 4.1.

TECHNOLOGY INVESTMENT DECISION MODEL

Stage 1: Basic Research

In Stage 1, fundamental scientific research for building and documenting core knowledge not tied to a specific defined need is evaluated, with the goal of generating new ideas. Objectives at this stage include identifying a new environmental technology or use of good science. Activities at this stage consist of basic laboratory experimentation, development of theory and analytical models, and proof of principle. The effectiveness of a project at Stage 1 is measured by whether it satisfies a subset of the programmatic driver criteria: specifically, technology and user need; technical merit; cost; and safety, health, environmental protection, and risk.

Gate 1: Entrance into Applied Research Stage

At this gate, projects addressing national interests and environmental performance needs enter the applied research stage. The technology developer or principal investigator (TD/PI) must address the programmatic driver criteria listed above.

APPENDIX A—DESCRIPTION OF OST'S TIDM

Stage 2: Applied Research

In the applied research stage, directed scientific or engineering research is conducted that has a link to remediation needs and results in a product concept. The goal is to conduct systems studies to address DOE priority needs. Research conducted includes proof-of-principle and lab-scale experimentation, with the objectives of defining data requirements, preparing experimental designs, determining material requirements, and determining business attributes. Project effectiveness at this stage is measured in terms of whether the project satisfies experimental design plan acceptance criteria and all of the programmatic driver criteria.

Gate 2: Entrance into Exploratory Development Stage

Gate 2 is a major decision point in the stage-gate model. At this gate, the TD/PI must show that the technology addresses a clearly defined DOE priority cleanup or waste management need and satisfies experimental design criteria. The TD/PI must also demonstrate knowledge of similar technology R&D activities taking place in other federal agencies, universities, industry, or international organizations to help information sharing, encourage cooperative relationships, and eliminate redundant research efforts. In addition, at Gate 2 the TD/PI initiates a comparison of the technology with the baseline and addresses the gate programmatic driver criteria.

Stage 3: Exploratory Development

The goal of the exploratory development stage is to conduct a systems study to address focus area priority needs. The technical feasibility of the project in terms of potential applications is evaluated (i.e., whether the technology can be developed sufficiently to solve the problem), with the objective of verifying that the concept can be linked to specific needs. Project activities at this stage includes laboratory-scale prototyping, analysis of user needs, estimates of life-cycle costs, and identification of functional performance requirements and operational concepts. The effectiveness of the project is measured by whether (1) it continues to satisfy experimental design plan acceptance criteria; (2) experimental performance meets program expectations; and (3) programmatic driver criteria are met.

Gate 3: Entrance into Advanced Development Stage

At this gate, the technology must be shown to be linked with clearly defined DOE or private-sector priority performance needs. Also, the TD/PI continues the baseline comparison and addresses gate programmatic driver criteria.

Stage 4: Advanced Development

The goal of Stage 4 is to show a specific DOE application of the product. A proof of design is required, and development includes full-scale laboratory testing, preliminary field testing, technical specification development, and infrastructure development plans. The objectives at this stage are assessment and validation of the technology's specifications and application by a review group. Effectiveness at this stage is measured by whether the application specifications satisfy the external review group's assessment, and whether programmatic driver criteria are met.

Gate 4: Entrance into Engineering Development Stage

Gate 4 is a major decision point, at which a review group completes an evaluation of information supplied by the focus area, TD/PI, and others to assess whether the technology is the right technology, in the right place, at the right time. The deliverables produced by the TD/PI address gate programmatic driver criteria and include a cost-benefit analysis showing the anticipated benefits of cost savings and risk avoidance, and strategies for DOE deployment, commercialization, cost sharing, regulatory compliance, and licensing. DOE's approval of expenditure at this gate depends on the commitment of an end user to implement the technology.

Stage 5: Engineering Development

At this stage, knowledge gained from R&D is used to develop systematically a detailed approach for full-scale design. The goal is classification of the technology as likely to exceed DOE baseline or to meet select government performance requirements or a problem set. Objectives at this stage include scaling up and refining detailed designs for prototypes and pilots, and clarifying the DOE deployment strategy and schedules to meet performance needs. This stage of development yields drawings, schematics, and computer codes; construction and demonstration units; prototypes and pilot-scale systems; system evaluation; reliability testing; infrastructure plans; and procurement specifications. Effectiveness is measured by the results of completed and documented preliminary tests, successful test plans, and satisfied programmatic driver criteria.

Gate 5: Entrance into Demonstration Stage

At Gate 5, the DOE deployment schedule is established. In addition, the TD/PI must address gate programmatic driver criteria, complete and document preliminary test results, and demonstrate that test plan requirements have been satisfied.

Stage 6: Demonstration

In Stage 6, the product or technology is subjected to a "real-world" demonstration, either at a DOE site or at another location, using actual waste streams and/or anticipated operating conditions with the goal of verifying design assumptions made up to this point. Objectives include conducting full-scale testing, system testing, and market conditioning to determine system suitability. Effectiveness is measured through programmatic driver criteria and acceptance of the technology by the end user.

Gate 6: Entrance into Implementation Stage

To pass through Gate 6, the results of the technology or system test must be fully documented. The technology partner must be fully invested (i.e., the procurement path is defined), and gate programmatic driver criteria must be engaged fully. In addition, implementation and commercialization viability must be defined clearly according to accepted business standards.

Stage 7: Implementation

At Stage 7, the product or technology has been proven to be viable, cost-effective, and applicable to required needs. The technology, if developed by

OST, is put into service by DOE and/or the end user and/or is transferred to the private sector. If not developed by OST, the technology is already commercially available. An end user signs a contract or approves operational use of the technology.

The TIDM incorporates several essential principles that DOE believes should be maintained:

1. Developers have to understand and address the needs and dynamics of the marketplace early in the innovation process.

2. Decision criteria must encompass both technical and nontechnical factors.

3. Formal decision points should provide the mechanism for determining investments in selected projects.

4. Decisions should reflect an EM R&D investment strategy.

Appendix B

Description of OST's Review Program[1]

The DOE–OST has developed an extensive review program with the goal of improving the design, management, and implementation of its programs and technical projects (see Figure B.1). OST seeks to integrate the results of its many reviews to aid managers in decision making. Results may feed into a number of decisions and activities, such as improving programs and management, formulating budgets, setting priorities, determining technology maturity and availability, and accelerating or decelerating programs. Reviews take place throughout the various levels of the EM organization and can be assigned to three different categories, described below:

    1. *Programmatic reviews* are designed to assess the appropriateness and effectiveness of the structure, goals, management, and budget of a particular organization or program. This type of review may be conducted internally (by OST staff) or externally (by standing bodies and groups with EM or OST oversight responsibilities). Reviewers must possess a high level of familiarity with the mission and activities of the organization or program being evaluated.

    2. *Technical assessment reviews* are designed to evaluate one or more projects with respect to organizational needs, objectives, responsibilities, or budget, and are used to provide the program or project manager with technical advice and direction. Evaluation criteria include effectiveness in contributing to meeting needs, cost-effectiveness, and public and regulatory acceptability.

---

[1] The material in this appendix is based on OST's description of its reviews provided to the committee in early 1997 (DOE, 1996), not on the committee's evaluation.

Technical assessment reviews are conducted internally. Reviewers include technology developers and users, who possess program and project knowledge expertise, and knowledgeable stakeholders, for example, regulators.

3. *Technical peer reviews*, also called merit reviews, are used to evaluate a project on its scientific or engineering basis, the competence of researchers, soundness of the research plan, and the likelihood of success. Reviewers may consider budgetary aspects of a project only from the standpoint of whether the proposed budget is adequate to complete the work. Technical peer reviews are conducted externally and independently, such that investigators whose work is being reviewed play no part in the selection or organization of the review panel. From the technical peer review, OST seeks to gain independent, unbiased, technical input or justification for funding project development. This type of review fits most closely the generally accepted definition of peer review.

These three types of reviews are performed across the organizational levels of the OST, which comprise department, program, and project levels. Reviews conducted at each of these levels are discussed below.

Department-Level Reviews

Department-level reviews assess the effectiveness of EM's technology development program and are all classified as programmatic reviews. These reviews are submitted to the Assistant Secretary for EM. OST considers department-level reviews to be external in that they are initiated and conducted independently of OST and the focus areas. The following groups conduct department-level reviews:

• *U.S. General Accounting Office.* GAO provides congressionally mandated programmatic oversight to DOE, and the observations and conclusions from GAO reports are made available to and used by OST.

• *National Research Council Committee on Environmental Management Technologies.* Until it was discontinued in September 1997, the CEMT[2] reviewed the broad issues of technology development, implementation, and evaluation within EM. It also evaluated specific technologies that EM considers most important toward reaching its goals. The CEMT submitted two annual reports to the Assistant Secretary for EM (NRC, 1995b, 1996). The

---

[2]Formerly the parent body of this committee.

APPENDIX B—OST'S REVIEW PROGRAM

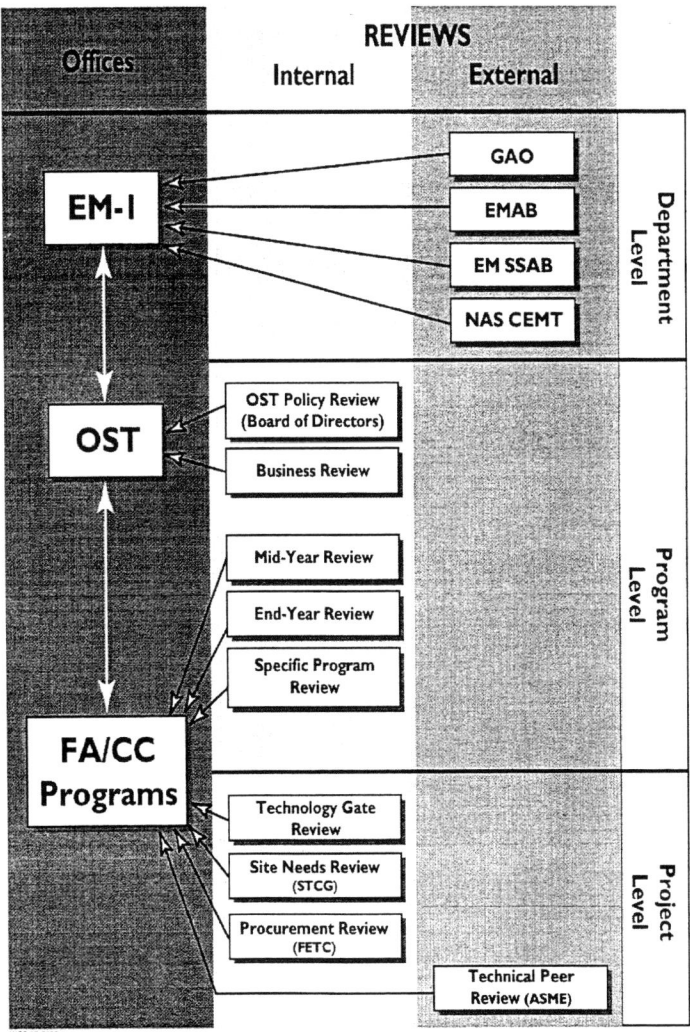

FIGURE B.1 Diagram of OST's review program showing different types of reviews, offices to which reviews are submitted, and the level of the organization at which reviews occur. NOTE: EM-1 = Office of the Assistant Secretary for Environmental Management; ER = Office of Energy Research; FETC = Federal Energy Technology Center; SSAB = Site-Specific Advisory Board; STCG = Site Technology Coordinating Group. SOURCE: DOE (1998b).

responsibilities of the CEMT have been assumed by six committees, including the present committee.

• *Environmental Management Advisory Board.* EMAB provides advice and recommendations to the Assistant Secretary for EM on a broad range of issues relevant to EM. EMAB is supported by EM, chartered under the Federal Advisory Committee Act (FACA), and composed of representatives from tribal, state, and local governments; other federal agencies; environmental and citizen groups; labor organizations; science organizations; and academia. Of the five subcommittees under EMAB, the Technology Development and Transfer Committee is most directly related to OST.

• *Environmental Management Site-Specific Advisory Board (SSAB).* The EM SSAB provides a means for community members to contribute to site-specific policy and technical decisions (e.g., future land use, integrated risk management, resource allocation, EM priority setting). The board is supported by EM, is chartered under FACA, and includes all board members from each local EM board throughout the DOE complex. Local boards have unique mission statements, operating procedures, and meeting schedules. Local board members are appointed by the Assistant Secretary for EM and include community members, members from local and tribal governments, and ex officio representatives from DOE, EPA, and state governments.

Program-Level Reviews

OST uses program-level reviews to chart progress being made and to assess the suitability of OST and focus area objectives, policies, and plans. These reviews are initiated and conducted within OST and the focus areas, and use both internal and external experts. Program-level reviews include the following:

• *OST Board of Directors Reviews.* The OST Board of Directors provides guidance and management for EM's science and technology programs, reviews entire program areas, and weighs high-level policy issues. The board consists of the Deputy Assistant Secretary for OST and the field office managers at the Savannah River, Idaho, and Hanford operations offices, and at the Federal (formerly Morgantown) Energy Technology Center (FETC). OST Board of Directors reviews are classified as programmatic reviews.

• *OST Business Reviews.* OST business reviews are conducted on a monthly basis and evaluate business management, program design, and execution of technology development activities, as presented by FA/CC program managers. The Deputy Assistant Secretary for OST performs these programmatic reviews.

- *Focus Area Midyear Reviews.* These reviews, conducted for each focus area, assess program management, direction, technical emphasis, and overall soundness of the respective focus area's technologydevelopment program. Both internal and external experts participate as reviewers, which include representatives of private industry, EM user groups, the scientific and academic communities, other federal agencies, and in some instances, members of the Community Leaders Network (OST's primary stakeholder organization). Midyear reviews are held in the second or third quarter of the fiscal year and are classified as programmatic reviews.
- *OST Year-End Review.* This internal review is an evaluation of individual projects in the context of overall program direction and available resources, and is intended to guide EM-50 headquarters staff as it finalizes programs for the budget year. In assessing projects, staff reviews associated technical peer review reports and considers progress made. The OST year-end review is classified as a technical assessment review.
- *Focus Area and Crosscutting Program Reviews.* These programmatic reviews assist managers in balancing the respective program's portfolio of technologies to fit the needs of users and stakeholders. Program reviews are conducted by the FA/CC program managers and are attended by technology users, stakeholders, and external industry and academic experts. The results of program reviews feed into EM's strategic planning.

Project-Level Reviews

Project-level reviews may be internal or external and are used to assess proposed or ongoing projects for scientific merit, potential for meeting a site need, potential for risk reduction and safety improvements, cost-effectiveness, regulatory and public acceptability, and commercial viability. Project-level reviews include the following:

- *American Society of Mechanical Engineers Peer Reviews.* ASME operates under a grant from DOE to conduct technical peer reviews of proposed or ongoing focus area projects. Reviewers are subject matter experts independent of DOE. Review panels are convened to review specific technologies or groups of technologies and are disbanded after issuing their reports. The ASME peer review process is described in further detail in Appendix A.
- *Site Technology Coordinating Group (STCG) Reviews.* STCG reviews are internal evaluations used to identify and prioritize site technology requirements. STCGs are located at each DOE operations office and are, in general, headed by the Technical Program Officer and composed of site

technology users, technology developers, site contractors, and stakeholders. The groups review focus area plans in order to ensure that technology development decisions address site cleanup needs. STCGs also serve as liaisons to regulators and stakeholders (including local SSABs), and ensure that their perspectives are incorporated into site technology decisions. The work of STCGs can be categorized as technical assessment reviews.

• *Procurement Reviews.* Procurement reviews are evaluations of proposals received in response to a solicitation for environmental restoration R&D projects. These solicitations are used when DOE is able to define, but unable to solve, a specific problem. Procurement reviews are conducted by FETC and are considered technical assessment reviews.

• *Environmental Management Science Program Reviews.* EMSP is an EM-sponsored research program designed to bridge the gap between fundamental research and needs-driven technology development for environmental restoration. Two types of reviews are conducted as part of the proposal evaluation process for EMSP. The Office of Energy Research (ER) conducts a peer review to evaluate the merit of proposals received and forwards selected proposals to a panel of EM managers. EM managers then conduct a technical assessment review to evaluate the proposals' relevance to EM remediation needs.

Appendix C

Biographical Sketches

C. Herb Ward (Chair) is the Foyt Family Chair of engineering at Rice University, where he directs the Energy and Environmental Systems Institute. Dr. Ward also directs the Department of Defense Advanced Applied Technology Demonstration Facility and for the past 15 years has directed the activities of the National Center for Ground-Water Research. In addition, Dr. Ward serves as codirector of the EPA-sponsored Hazardous Substances Research Center/South & Southwest. His research interests include the microbial ecology of hazardous waste sites, biodegradation by natural microbial populations, microbial processes for aquifer restoration, and microbial transport and fate. Dr. Ward is also an expert on the technical issues surrounding cleanup of the nation's nuclear weapons complex. He has served on several NRC committees, including the Committee on Environmental Management Technologies. Dr. Ward received his Ph.D. in plant pathology, genetics, and physiology from Cornell University and an M.P.H. in environmental health from the University of Texas.

Barry Bozeman is the director of the School of Public Policy at the Georgia Institute of Technology, where he specializes in science and technology policy, focusing on research and development impact evaluation, technology transfer, and commercialization. Dr. Bozeman is an expert on the use of peer review to evaluate the impacts of research and development. His work on peer review has included a state-of-the-art review paper titled "Peer Review and Evaluation of R&D Impacts" (in *Evaluating R&D Impacts: Methods and Practice*, Bozeman and Melkers, 1993), as well as studies on peer review for the National Science Foundation and the U.S. Air Force. He has served on the NRC's Committee to Address Continued Review of the Tax System's Modernization of the Internal Revenue Service. Dr. Bozeman received his Ph.D. in Political Science/Public Administration from the Ohio State University.

Radford Byerly, Jr. recently retired as vice-president for public policy of the University Corporation for Atmospheric Research after a distinguished career in academia and government, specializing in science management and policy. Among his many positions, Dr. Byerly worked at the National Institute of Standards and Technology (then the National Bureau of Standards) in the environmental measurement and fire research programs; served as chief of staff of the U.S. House of Representatives Committee on Science and Technology; and was director of the University of Colorado's Center for Space and Geosciences Policy. He currently serves as a member of NASA's Space Science Advisory Committee and serves on NSF site visit committees and review panels. Dr. Byerly is a member of the NRC Board on Assessment of NIST. He received his Ph.D. in physics from Rice University.

Linda Capuano is the vice-president of technology and new business development at AlliedSignal Aerospace. In this capacity, Dr. Capuano is responsible for restructuring the company's review process for selecting research and development programs. Prior to joining AlliedSignal, she worked as vice-president of business development at Conductus and held a number of engineering and management positions at IBM. Dr. Capuano also served on the DOE Task Force on Alternative Futures for the DOE National Laboratories ("Galvin task force"), which explored the relevance of national laboratory research, including the role of peer review in research at DOE's national laboratories. Dr. Capuano received her Ph.D. in materials science from Stanford University.

Richard Conway is a recently retired senior corporate fellow at Union Carbide Corporation. His areas of expertise include contaminated site remediation, hazardous waste management, and environmental risk analysis of chemical products. Mr. Conway was elected to the National Academy of Engineering in 1986 for his contributions to environmental engineering and for the development of improved treatment processes for industrial wastes. He has received many awards and honors, including the Hering Medal, Gascoigne Medal, Dudley Medal, Rudolfs Medal, and Rachel Carson Award. Mr. Conway has been involved in numerous NRC activities, including the Board on Environmental Studies and Toxicology, Water Science and Technology Board, Committee on Peer Review of Department of Defense Environmental Scholarships and Grants, and Commission on Engineering and Technical Systems. He earned his M.S. in environmental engineering at the Massachusetts Institute of Technology.

APPENDIX D—BIOGRAPHICAL SKETCHES 111

Thomas Cotton is vice-president of JK Research Associates, Inc., where he is a principal in activities related to radioactive waste management policy and strategic planning. Before joining JK Research Associates, he dealt with energy policy and radioactive waste management issues as an analyst and project director during 11 years with the U.S. Congress Office of Technology Assessment. His expertise is in public policy analysis, nuclear waste management, and strategic planning. Dr. Cotton has served the NRC as a member of the CEMT and the Committee on the Remediation of Buried and Tank Wastes. He received a Ph.D. in engineering-economic systems from Stanford University.

Frank Crimi recently retired as vice-president for Lockheed Martin's Advanced Environmental Systems Company. He joined Lockheed in 1992 after completing 34 years in engineering and management positions with the General Electric Company. Mr. Crimi has more than 30 years experience in the design, operation, and maintenance of DOE naval nuclear power plants, with special emphasis on decontamination and decommissioning of nuclear facilities. He was General Electric project manager for the Shippingport Atomic Power Station decommissioning and recently chaired the Long Island Power Authority's Independent Review Panel during decommissioning of the Shoreham Nuclear Power Station. Mr. Crimi was a member of Public Service of Colorado's Management Oversight Committee for the Fort Saint Vrain Nuclear Generating Station decommissioning. He currently is on advisory boards for the decommissioning of the Trojan and Connecticut Yankee Nuclear Power Plants. Mr. Crimi completed a B.S. in mechanical engineering at Ohio University in Athens and did graduate studies in mechanical engineering at Union College, Schenectady, New York.

John Fountain is a professor of geochemistry at the State University of New York at Buffalo. His research focuses on various aspects of contaminant hydrology, including aquifer remediation and the characterization of fractured rock aquifers. Dr. Fountain is a member of the NRC's Committee on Technologies for Cleanup of Subsurface Contaminants in the DOE Weapons Complex. He received his Ph.D. in geology from the University of California, Santa Barbara.

David T. Kingsbury is vice-president and chief information officer at Chiron Corporation. His research interests include computational biology and databases, molecular diagnostic techniques in medical microbiology, and the biochemistry and mechanisms of pathogenesis of the slow (unconventional) viruses. Dr. Kingsbury is an expert on the administration of peer reviews, having served for four years as assistant director for biological, behavioral, and social sciences at the National Science Foundation. He also serves as editor-in-chief of the *Journal of Computational Biology* and is North American editor of the *Journal of Chemical Technology and Biotechnology*. Dr. Kingsbury received his Ph.D. in biology at the University of California, San Diego.

Gareth Thomas is a professor in the Graduate School of the Department of Materials Science and Mineral Engineering at the University of California, Berkeley. He is a renowned expert in the theory and application of electron diffraction and high-resolution microscopy to problems in materials science and engineering. Dr. Thomas is a member of both the National Academy of Sciences and the National Academy of Engineering. In addition to holding a variety of positions at Berkeley, he also has held positions at the Lawrence Berkeley Laboratory, where he founded the National Center for Electron Microscopy. Dr. Thomas has been involved in peer reviews for a variety of scientific societies and scholarly journals and currently serves as editor-in-chief for the journal *Acta/Scripta Materialia*. He has served on the NRC Committee on Materials Research Opportunities and Needs in Materials Science and Engineering. Dr. Thomas received his Ph.D. and Sc.D. from Cambridge University.

Appendix D

Acronyms

AIBS	American Institute of Biological Sciences
ASME	American Society of Mechanical Engineers
ASTD	Accelerated Site Technology Deployment
BES	Office of Basic Energy Sciences (DOE)
CEMT	Committee on Environmental Management Technologies (NRC)
D&D	decontamination and decommissioning
DOD	U.S. Department of Defense
DOE	U.S. Department of Energy
EM	Office of Environmental Management (DOE)
EMAB	Environmental Management Advisory Board (DOE)
EMSP	Environmental Management Science Program (DOE)
EP	Executive Panel of ASME Peer Review Committee
EPA	U.S. Environmental Protection Agency
ER	Office of Energy Research (DOE)
ESTCP	Environmental Security Technology Certification Program (DOD)
FACA	Federal Advisory Committee Act
FA/CC	Focus area/crosscutting area (DOE)
FETC	Federal Energy Technology Center
FY	fiscal year
GAO	U.S. General Accounting Office
GPRA	Government Performance and Results Act
IDI	Information Dynamics, Incorporated
NASA	National Aeronautics and Space Administration
NGO	Non-governmental organizations
NIH	National Institute of Health

NIST	National Institute of Standards and Technology
NRC	National Research Council
NSF	National Science Foundation
OLMSA	Office of Life and Microgravity Sciences and Applications (NASA)
OST	Office of Science and Technology (DOE)
PI	principal investigator
PRC	Peer Review Committee (ASME)
R&D	research and development
RSI	Institute for Regulatory Science
SAB	Science Advisory Board (DOD)
SBIR	Small Business Innovation Research
SERDP	Strategic Environmental Research and Development Program (DOD)
SON	Statement of Needs
SSAB	Site-Specific Advisory Board (DOE)
STCG	Site Technology Coordinating Group (DOE)
TD	Technology Developer
TDI	Technology Deployment Initiative
TIDM	Technology Investment Decision Model
TTAWG	Technology Thrust Area Working Group (DOD SERDP)
USNRC	U.S. Nuclear Regulatory Commission
WES	Waterways Experiment Station (U.S. Army Corps of Engineers)